Mathematics Teaching Practice: Guide for university and college lecturers

'Talking of education, people have now a-days' (said he) 'got a strange opinion that every thing should be taught by lectures. Now, I cannot see that lectures can do so much good as reading the books from which the lectures are taken. I know nothing that can be best taught by lectures, except where lectures are to be shewn. You may teach chymestry by lectures – you might teach making of shoes by lectures!'

James Boswell: The Life of Samuel Johnston, LLD, 1766

'Mathematics possesses not only the truth, but supreme beauty – a beauty cold and austere like that of stern perfection, such as only great art can show.'

Bertrand Russell: The Principles of Mathematics, 1902

About John Mason

John Mason has been teaching mathematics ever since he was asked to tutor a fellow student when he was aged only fifteen. In college he was first an unofficial tutor, then later an official tutor for mathematics students in the years below him, and also found the time to tutor school students as well. He began his university career in Toronto, receiving first a BSc in Mathematics from Trinity College, and then an MSc while at Massey College. He then studied for a PhD in Combinatorial Geometry in Madison, Wisconsin, where he encountered Polya's film *Let Us Teach Guessing*. Seeing the film evoked a style of teaching he had first experienced at high school from his mathematics teacher, Geoff Steel, and his teaching changed overnight.

He then took up an appointment at the Open University, becoming involved, among other things, in the design and implementation of the first mathematics summer school (5000 students over 11 weeks on three sites simultaneously). Drawing upon his own experiences as a student, he created active-problem-solving sessions, which later became investigations. He also developed the idea of project-work for students in their second year of pure mathematics. In 1982 he wrote *Thinking Mathematically* with Leone Burton and Kaye Stacey, a classic that has been translated into four languages and is still in use in many countries around the world. It has been used with advanced high school students, with graduates becoming school teachers, and with undergraduates who are being invited to think about the nature of doing and learning mathematics. He is also the author of *Learning and Doing Mathematics*, which was originally written for Open University students, then modified for students entering university generally.

At the Open University he led the Centre for Mathematics Education in various capacities for fifteen years, during which time it produced the influential *Routes-to Roots-of Algebra* and numerous collections of materials for teachers at every level. His principal focus is thinking about mathematical problems, and supporting others who wish to foster and sustain their own thinking and the thinking of others. Other interests include the study of how authors have expressed to students their awareness of generality, especially in textbooks on the boundary between arithmetic and algebra, and ways of working on and with mental imagery in teaching mathematics. The contents of this book spring from a lifetime of collecting tactics and frameworks for informing the teaching of mathematics. Along the way he has articulated a way of working, developed at the Centre, that provides methods and an epistemologically well founded basis for practitioners to develop their own practice, and to turn that into research.

Mathematics Teaching Practice: Guide for university and college lecturers

John H. Mason, BSc, MSc, PhD
Centre for Mathematics Education
Open University
Milton Keynes, UK

WOODHEAD PUBLISHING

in association with

The Open University

Oxford Cambridge Philadelphia New Delhi

Published by Woodhead Publishing Limited,
80 High Street, Sawston, Cambridge CB22 3HJ
www.woodheadpublishing.com

Woodhead Publishing, 1518 Walnut Street, Suite 1100, Philadelphia,
PA 19102-3406, USA

Woodhead Publishing India Private Limited, G-2, Vardaan House, 7/28 Ansari Road,
Daryaganj, New Delhi – 110002, India
www.woodheadpublishingindia.com

First published by Horwood Publishing Limited, 2002; reprinted 2003. 2004
Reprinted by Woodhead Publishing Limited, 2011

British Library Cataloguing in Publication Data
A catalogue record for this book is available from the British Library

ISBN 978-1-898563-79-2

Preface

'The mathematical background of our undergraduates is undermining the quality of their degree.'
(Sutherland and Dewhurst, 1999, p6)

It is vital, in our increasingly technological society, that a wide range of people have positive experiences of mathematics, developing confidence both in using what they do know and in finding out what they do not know when they need it. Mathematics lies at the heart of many different disciplines and, whether they are taught by mathematicians or by experts in other disciplines, all students of mathematics need more than simply to master mysterious manipulative techniques. This book maintains that all students can do more, and aims to show how it can be done.

Whoever does the teaching, it is vital to encourage students to engage with mathematical thinking, because otherwise they may be reduced to trying to remember and use formulae and techniques which may not be appropriate to their situation or, worse, may try to avoid using mathematics at all cost. To assist teachers of mathematics from whatever background, this book:

- provides a collection of useful practices for the teaching of mathematics in colleges and universities;
- indicates some aspects of mathematics which are worth bearing in mind or being aware of while preparing, conducting, and reflecting upon sessions;
- suggests ways of thinking about ever-present tensions in teaching.

The aim is to help anybody teaching mathematics who suspects that much more is possible than was done for them, and to show that it is possible to teach so that learning is both effective and efficient, even pleasurable.

This book is written from the perspective that mathematics has to be learned through actively engaging with it. This means not only actively making sense of definitions, theorems, and proofs, but also participating in other aspects of mathematical thinking such as specialising and generalising, conjecturing and convincing, imagining and expressing, organising and classifying, and through posing and resolving problems.

Furthermore, students need to manipulate 'things' that inspire their confidence in order to begin to make sense of generalisations provided by a textbook or their lecturer, and eventually bring these to articulation both in their own words and in formal terms. Learning mathematics is not a monotonically smooth process; it frequently requires going back over old ground to see it from a fresh perspective, reformulating concepts and ideas in new and often more precise terms.

The suggestions made in this book are based on this perspective, but the suggestions will be of use no matter what your perspective on mathematics and how it is most effectively learned and taught.

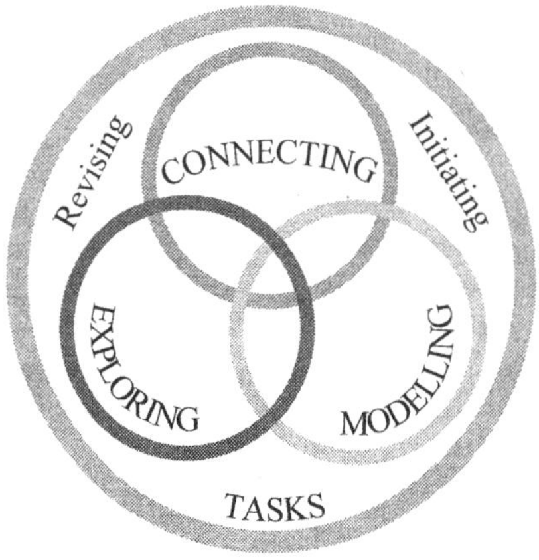

The book begins with descriptions and partial diagnoses of some classic student difficulties, and descriptions of possible actions that might improve the situation. Subsequent sections then address the principal modes of interaction: lecturing, tutoring, task construction, and assessment. Throughout, the aim is to stimulate students to take the initiative in working on mathematics, rather than just sitting and responding passively to what is presented to them.

The underlying theme is expressed in the image of interlocking rings, which suggest an interweaving of exploration, modelling and connection forming as purposes for tasks given to students. These can be used both to initiate and to revise or review a topic.

Outline

This book is intended to support you in developing and extending your range of practices. It could also serve as the basis for building a portfolio of evidence of professional development. It is certainly not intended to be read from cover to cover. Rather, it is intended as a cross-referenced resource to call upon when you want some fresh ideas, or when some aspect of your teaching is not going quite as smoothly as you might wish.

If you have recently started teaching, you may wish to concentrate on two tactics selected from Chapter 1, and the tactic *Being Mathematical* from Chapter 3.

Chapter 1 is built around a collection of common student mistakes and misconceptions, and suggests partial diagnoses and useful tactics for dealing with them. Chapter 2 is devoted to lecturing, and Chapter 3 to tutoring. Chapter 4 is concerned with constructing tasks for a variety of different purposes, including assessment, while Chapter 5 focuses on marking. Chapter 6 then considers the role of history in teaching mathematics, while Chapter 7 summarises the previous chapters by raising some endemic tensions and issues in teaching mathematics, and suggesting ways of addressing them. Finally, there are two appendices: a representative collection of challenging explorations for first year undergraduate mathematics students in Appendix A; and, in Appendix B, an example of the unfolding of a particular topic according to some of the suggested framework structures. The last appendix is intended as a form of 'worked example' of how to prepare to teach a topic, in this case convergence of series in which all terms are non-negative.

The text as a whole forms a richly interconnected web of tactics and sensitivities. Consequently, the same ideas arise in different sections, though sometimes described using slightly different vocabulary.

Effective Teaching

Teaching well requires expertise different from that required to be a creative mathematician. Whereas experts draw their colleagues into their own world of discovery and creation, and expect their audience to be able to follow their arguments and insights, teaching students requires more than this. Not only do you have to inspire novices and draw them into your world, through being what Philip Davis calls 'the sage on the stage', but you also have to stand by and support them while they work on, struggle with, and reconstruct ideas for themselves, acting as 'a guide on the side'. Furthermore, to be really effective in supporting them, you have to be able to enter their world and remain within it, as this is the only way to really appreciate what they are struggling with. Students are people after all, with hopes and fears, strengths and weaknesses, propensities and habits. Most of them need assistance in undertaking the mental actions that experts find intuitive and natural.

Although creativity in mathematics and in mathematics education are very different in form and function, they are interwoven components of a tapestry. Working on your teaching develops your awareness of and sensitivity to the structure, history, and pedagogic implications of the mathematical topics that you teach. That awareness and sensitivity can also inform your research practice, as well as revealing topics for further research, both in mathematics and in mathematics education.

Structural Summary

Recognising the natural desire of mathematicians to be told the essential structure without a lot of words, while also acknowledging that developing one's teaching is a long slow process, I offer a brief structural summary. As with mathematical exposition, this summary may not make a great deal of sense now, but I hope it will attract you to read further. Of course, the best kind of summary is one that you reconstruct for yourself, just as you only really understand a theorem or a technique when you can reconstruct it for yourself when needed.

There are six main modes of interaction between student, content, and tutor:

- *Expounding*, or attracting your students into your world of experience, connections, and structure;
- *Explaining*, or entering the world of the student and working within it;
- *Exploring*, or guiding your students in fruitful directions as they sort out details and experience connections for themselves;
- *Examining*, when students validate their own developing criteria for whether they have understood, by submitting themselves for assessment;
- *Exercising*, when students are moved to rehearse techniques and to review connections between theorems, definitions and ideas;
- *Expressing*, when students are moved to express some insight.

All six modes contribute to effective learning, so effective teaching employs them all.

It is natural for students to struggle with new ideas, but that struggle is only productive when they learn from the experience, and become aware of their innate powers to think mathematically. These powers include:

- *Imagining* and *Expressing*;
- *Specialising* (particularising) and *Generalising* (abstracting);
- *Conjecturing* and *Convincing* (yourself, a friend, and then a reasonable sceptic);
- *Ordering*, *Classifying*, and *Characterising*.

The effect of using these powers is to develop the ability to perceive and think mathematically: to notice opportunities to ask mathematical questions as well as to explore and investigate; to make sense of both the material and mathematical world; and to recognise connections between apparently disparate topics. It is a process of altering the structure of attention, of altering what is attended to and how that attention is configured. Teaching mathematics is a matter of both educating your students' awareness and training their behaviour, by harnessing their emotions.

Mathematics is much more than a collection of techniques for getting answers, and much more than a collection of definitions, theorems and proofs. It is a richly woven fabric of connections. Many of those connections can be revealed by becoming sensitised to underlying mathematical themes such as:

- Doing and Undoing;
- Invariance Amid Change;
- Freedom and Constraint;
- Extending Meaning.

The aim of this book is to present a variety of tactics that may provide the means for achieving these aims, through becoming more aware of opportunities to act in ways that stimulate your students to take the initiative.

Acknowledgements

A work like this is the product of many collaborations, not all of them witting! I am grateful to the many colleagues whom I have watched teach, or whose teaching I have heard described. Most of the proposals I have tried out myself in some form or other. I am particularly grateful to my late colleague Christine Shiu for continued encouragement to undertake this project. I also owe a great debt to Liz Bills, Bob Burn, Dave Hewitt, Eric Love, Elena Nardi, Peter Neumann, Graham Read, Dick Tahta, and Anne Watson for their detailed and insightful suggestions and support at various times.

Contents

Preface v
Outline vi
Effective Teaching vii
Structural Summary vii
Acknowledgements viii

0 Some General Opening Remarks **1**
Preparing to Teach 2
Reflection 5

1 Student Difficulties with Mathematics **7**
Introduction 7
Difficulties with Techniques 10
Difficulties with Concepts 17
Difficulties with Logic 27
Difficulties with Studying 29
Difficulties with Non-routine Problems 33
Difficulties with Applications 34
Reflection 38

2 Lecturing **39**
Introduction 39
Lecture Structure 40
Employing Screens 41
Tactics 47
Other Lecturing Issues 58
Reflection 69

3 Tutoring **71**
Introduction 71
Conjecturing Atmosphere 72
Scientific Debate 73
Asking Students Questions 75
Getting Students to Ask Questions 76
Worked Examples 77
Assent – Assert 82
Collaboration Between Students 86
General Tactics 87
Advising Students How to Study 95
Structuring Tutorials 100
Reflection 103

4	**Constructing Tasks**	**105**
Introduction		105
Purposes, Aims, and Intentions		105
Different Tasks for Different Purposes		108
Forms in Mathematics		124
Learning Outcomes		127
Inner and Outer Tasks		130
Student Propensities		132
Boundary Examples		135
Reflection		140
5	**Marking and Commenting**	**141**
Introduction		141
Allocating Marks		141
Feedback Among Markers		143
Feedback to Students		145
Feedback from Students		153
Reflection		153
6	**Making Use of History**	**155**
Introduction		155
Why Use History?		155
How to Use History		157
Reflection		160
7	**Issues and Concerns in Teaching Mathematics**	**161**
Introduction		161
Tensions		163
Issues		173
Resources Upon Which to Call		184
Frameworks for Informing Teaching		187
Reflection		194
Appendix A:	**Exploratory Tasks**	**195**
Appendix B:	**Convergence Case Study**	**197**
Salient Items		197
Great examples		198
Concept Image Framework		200
Bibliography		**203**
Historical Sources and Resources		209
Teaching of Specific Mathematical Topics		210
Classroom Techniques		211

Research into Teaching Mathematics at Tertiary Level 212
Perspectives on Mathematics 212
Websites 212

Index of Tactics, Issues, Themes, Frameworks and Tensions **215**
Tactics 215
Issues 218
Themes 218
Frameworks 218
Tensions 218

initiative to act upon the mathematical ideas and make sense of them, and not just attempt to master a succession of techniques.

Some colleagues have commented that the huge range of tactics suggested here makes them feel guilty that they do not spend as much time on their teaching as perhaps they could. My own experience has been that time spent working on teaching has enhanced my research. Although this may seem unlikely, the principal effect of teaching has been to sharpen my sensitivity to my own thinking processes, to where my attention is directed. When this carries over into research, it enhances research activity as well. In addition, interesting connections and problems can come to light when constructing tasks for students (Cuoco, 2000). However, you must test this conjecture, like everything else suggested here, for yourself. It is also unwise to work on more than two aspects of your teaching at any one time. A wide range of possibilities is offered here, in the hope that there will be something to suit everyone, whatever their teaching task. There is no implicit or explicit suggestion to work on everything!

Preparing to Teach

Before embarking on this journey, it may be useful to ponder your basic assumptions and beliefs about how mathematics is most effectively learned, because these determine to a large extent what sorts of things you will try and why you might want to try them. Just as with assumptions in mathematics that need to be explicitly stated so as to be taken into account in proofs, it is valuable to bring your assumptions and beliefs about teaching and learning to the surface so that they too can be examined, questioned, and perhaps modified. Like any mathematical conjecture, as long as they remain implicit or below the surface of your awareness, they will have a strong influence but will not be open to challenge.

Task: Assumptions and Beliefs

Mark the entries in the following table that, together, come closest to capturing what you feel to be the most important aspects of learning and teaching mathematics, adding your own if you wish.

Students learn mathematics most effectively by		
Doing lots of examples for themselves	Reading through their notes or a textbook carefully in the light of their own examples	Reconstructing theorems and techniques for themselves
Being shown how ideas can be formalised or abstracted	Following a clearly laid out presentation, line by line and symbol by symbol	Following the development of definitions, lemmas, theorems, and proofs
Posing and solving problems	Working by themselves	Working with others
	Discussing topics with others	

Perhaps your students do not really know what it is like to be mathematical, to think mathematically, to recognise situations as opportunities to ask questions that can be worked on mathematically. Maybe what your students need most is to be in the presence of someone who is 'being mathematical': someone who asks mathematical questions mathematically (Mason *et al.*, 1982, Mason, 2000). In almost all cases the topics being taught are in fact well rehearsed and familiar to the lecturer, so it is tempting to act like a talking textbook, to reproduce the distilled essence without treating the topic as an example of how mathematicians work and think. Yet if we want our students to be enthused by mathematics, to approach it eagerly and positively, and if we want them to appreciate what mathematics is like as a discipline rather than simply as a body of definitions, theorems, proofs and techniques, then it behoves us to be mathematical with and in front of our students. If we want our students to encounter not just techniques, but structures, heuristics, and ways of thinking pertinent to the particular mathematical field being taught, then we need to display these explicitly.

This does *not* mean that it is effective to walk in and solve a lot of problems, formulate definitions and prove theorems in front of them, mindless of their presence. On the other hand, neither is it effective to give a truncated and stylised presentation which supports the impression that mathematics is completely cut, dried and salted away, that it is something that one can either pick up easily or not at all. The most effective method is to display aspects of mathematical thinking, such as forming and questioning mental images supported by diagrams, constructing examples to probe as well as illustrate theorems and techniques, asking mathematical questions about situations, and making and modifying conjectures publicly.

It must be noted that in our present consumer-oriented society, student comments on their experience as students are of great importance for quality assurance. Consequently, if you are going to embark on changing your practices, you must make sure you take your students with you. One useful way to think about this is in terms of the existence of an implicit contract between you and your students: they expect you to give them facts and tasks, and they expect that, through memorising the facts and doing the tasks, learning will take place. You, in turn, expect them not only to do the tasks, but to be able to reconstruct the techniques and use these in a variety of contexts. This is one of a number of issues that have been studied and debated in some depth in the mathematics education literature and which are discussed in Chapter 7.

Comment: In looking through this list of short statements, you probably found something positive in most of them. Try to find or formulate one or two that sum up for you how students learn best; the sort of thing you might find yourself saying to a colleague, or thinking when listening to someone discussing various forms of teaching.

It is likely that you drew on your own experience, perhaps recognising elements of what seemed to work for you. However, memories are not always entirely reliable, though we often tend to base our teaching on what we think we did as students.

Task: Teaching

Now consider your responsibilities as a teacher, again adding your own entries as necessary.

My responsibility as a teacher is to		
Set out all the details as clearly and logically as possible	Stimulate my students to making sense of something for themselves	Provide motivation and applications for the material covered in the text
Cover all the techniques they will be tested on	Show my students how to wrestle with mathematics the way a mathematician does	Startle and surprise my students, generating dilemmas that they will have to resolve
Concentrate on techniques	Concentrate on meaning and understanding	Introduce new ideas and concepts
	Display links between topics	

Now go back and see in what way you disagree with, or place much less emphasis on, the other entries (each entry is espoused by some very good teachers!).

Comment: As before, you probably found at least something positive in most of the statements. Try to again find or formulate the one or two that represent for you the principal contribution you can make to supporting your students in learning effectively.

Did you think to add something about assessment, such as that final assessment should be similar to the assessments and exercises used during the course, or that assessment needs to be both challenging and yet confidence developing?

No matter what perspective you hold, perhaps the fundamental question for any lecturer, tutor, or marker is how to stimulate their students to take the initiative. Perhaps your students do not actually know what it means to take the initiative with respect to mathematics. After all, their experience may be of a complex subject with a multiplicity of technical terms and techniques, which all seem well worked out. They may see their task as being to reproduce set behaviour under examination conditions.

Reflection

Task: Reflection

What specific issues about teaching mathematics trouble you at the moment?

What examples can you find from your own experience to illustrate or challenge the beliefs outlined briefly above?

Comment: Issues mentioned by current lecturers include:

- how to interest your students in really working at mathematics, especially in service courses;
- how to leave a lecture feeling that some students have actually got something from it;
- how to help students in a tutorial or problem class without just doing it for them;
- how to cope with a wide range of backgrounds and mathematical facilities.

1 Student Difficulties with Mathematics

Have you ever encountered students who make extraordinary errors in the midst of supposedly routine calculations?

Have you ever encountered students who seem not to remember what they were taught last year, last term, or even last week?

Have you ever encountered students who score well in tests on routine tasks but do not seem to think of using those techniques in other contexts?

Have you ever encountered students who seem to understand (they pass tests) but who complain about not understanding?

Introduction

Thinking about mathematics for oneself is very different from thinking about teaching mathematics to others, and communicating mathematics to someone else is not nearly as transparent a process as it might at first seem. While the speaker or author has a deep sense of the connections between topics or theorems, some of which come to the surface and are articulated, the reader or listener has to infer the presence of those connections from the fragments that they actually hear, see, and can make some sense of. You might hope that every student will hear every word and see every symbol and diagram, but attention is a curious power, able to cut off outside sounds while inner thoughts proceed.

Even when an author or a teacher has clearly laid out every definition, lemma, theorem, proof, and context of application, complete with motivating, illustrative and typical worked examples, the student is still faced with making sense of the mass of material. Just because you have clearly labelled something as an example and something else as a theorem, it does not follow that all students will make the same distinctions, or, if they do, make them in the same way or with the same significance. What seems clear to the teacher may be confusing to a student who does not make the same distinctions, does not stress the same points as the author. A lecturer who works through an example step by step probably sees it as but a special case of a general technique, while the student, unaware of the general technique, tries to make sense of each step of the particular example.

Students have to distinguish the various components (definitions, theorems, proofs, examples, contexts, motivation, worked examples) and then use them to, in effect, reconstruct the topic for themselves. In other words, teaching mathematics is not simply a matter of telling the students how to do technical manipulations, and hoping that they will learn to do those with facility and know when and how to use them in other contexts. In order to reach such a state, each student has to reconstruct the topic, the connections, and the techniques for themselves, using what they have encountered already as a guide. If they cannot do this, they have to memorise everything, or hope that familiarity achieved through repetition will somehow transform itself

Difficulties with Techniques

Here are three examples of the kinds of difficulties that tutors report.

Difficulty T1: Algebraic Manipulation	*Diagnoses*
Students are uncertain about negative signs in expressions such as $-2(x-3) = -2x-6$. Students make errors when working mentally: they solve $6x+3=0$ to obtain $x=1/2$ or $x=-2$; their solution to $x^2 = 4x$ is $x=4$ only, missing out $x=0$. Students cancel inappropriately: they write $2\cancel{h}^2/\cancel{h} = 2^2 = 4$, or $(\cancel{x}^2+y)/\cancel{x} = x+y$. Students assume linearity: they write $(a+b)^2 = a^2+b^2$, $1/a+1/b = 1/(a+b)$, or $\ln(a+b) = \ln a + \ln b$, usually in some disguised form; the same applies to $\sin(a+b)$ and e^{a+b} . Students confuse similar notation, for example $\sin^{-1} x$ and $(\sin x)^{-1}$.	Students *may* know better, and be making a slip; they may be caught up in working on a larger problem and perform an automatic, unheeded, incorrect act. They need to awaken their inner 'checker'. Students *may* genuinely think their version is correct. Students *may* not be aware of there being a difference between what they do and what is correct. They may not have stopped to ask themselves if there is a difference or an alternative.
Comment: Unless some time is spent sorting out these confusions they are likely to persist, but there is often insufficient time to sort them out and yet still cover the syllabus. Thus arises a fundamental tension both for teachers and for students: the trade-off between coverage and pace (see *Issue: Time – Coverage and Pace*, p164). The suggestions in this book are intended to provide ways of reducing this tension.	

Tactic: Using Common Errors

Tactic:

Collect common errors that you see (or have seen) students make on assignments. Colleagues may have a few to hand as well. You can then set these as a 'test' near the beginning of the course (for example, 'Find the mistake in …') though it might put some students off if it is too extensive. Alternatively, you can hand out a sheet of classic errors, and address one in each lecture (briefly, say in the first 3 minutes). (See the *Common Mathematical Errors* website.)

In a tutorial, ask students to work in pairs to explain what is wrong with one (or more) of them, before explaining it to the whole group (see *Tactic: Talking in Pairs,* p49). You could establish a practice in which the first activity in each tutorial is to work on one common error in this way, for no more than 3 to 5 minutes. You may even find that students start to bring their own errors to tutorials and ask for assistance.

The idea is to expose students to the possible confusion, with the expectation that having been awakened to it, they will be more alert in the future. Some lecturers have found this to be very effective.

Tactic: Specialising, Generalising and Counter-examples

Students arrive with the ability to specialise a general statement, and to detect patterns in particular examples and then to re-express these in a more general way. They can use these powers to work out what is common to a collection of exercises, and to construct examples to illustrate theorems. Indeed, if they cannot do this, then perhaps they do not fully appreciate or understand the theorem. The importance of appreciating generality has been recognised since ancient times.

'... "man has a wisdom of analogy" that is to say, after understanding a particular line of argument one can infer various kinds of similar reasoning, or in other words, by asking one question one can reach ten thousand things. When one can draw inferences about other cases from one instance and one is able to generalize, then one can say that one really knows how to calculate .... The method of learning: after you have learnt something, beware that what you have learnt is not wide and after you have learnt widely, beware that you have not specialized enough. After specializing you should worry lest you do not have the ability to generalize. So by having people learn similar things and observe similar situations one can find out who is intelligent and who is not. To be able to deduce and then to generalize, that is the mark of an intelligent man ... If you cannot generalize you have not learnt well enough ...'
(Zhōubì Suànjìng, quoted in Li and Dù, 1987, p28)

However, if generalities are always stated and particularised for them, students may not think of generalising and particularising as something that they are supposed to do. They may not associate it with mathematical thinking. To counteract this, students can, for example, be encouraged to make up examples and to follow the argument of a theorem through their own example, with a view to working out why the proof works, and what the theorem is actually saying.

It is useful to be explicit about using these techniques yourself, and to call explicitly upon students to use them as well.

Tactic: Be explicit to students about when *you* are particularising or exemplifying, and when you are stating a generality.

For example, when you announce a theorem, publicly acknowledge your construction of an example to 'see if it works'. Offer three different examples, and publicly draw attention to the features which make them all similar and which are captured in the theorem. With a sheet of errors, students can be asked to locate three 'counter-examples' to each error, and perhaps even to resolve the question of under what conditions the 'error' is actually correct, if any.

Tactic: Instead of always stating a generality and then offering one or more worked examples, try starting with the particular examples and then inviting students to express what they see as common between them. Also, try getting them to particularise a generality for themselves, in order to help them to appreciate what the generality entails.

The aim is *not* to induce students to 'learn' different notations, but rather to provide practice in accommodating to changes in notation, which is essential for when they read different texts or encounter the same ideas in different courses. To prepare them for this flexibility, the idea is to stimulate students to use different notation to express ideas for *themselves*, and to become aware that there are different ways of thinking about the same concept. This means that, when they encounter an unfamiliar expression, they have the tools to decode and re-express what is being said in their own way.

Tactic: Boundary Examples

Tactic: As part of a review of a topic, ask students to offer two or three 'examples' of a mathematical object satisfying a particular definition or theorem, with increasingly complex properties. Ask them to make sure that, at each stage, their examples *do not* satisfy the condition which follows.

For example:

- write down a function specified on $[0,1]$;
- write down a function which is also continuous on $[0,1]$;
- write down a function which is also differentiable on $[0,1]$;
- write down a function which also has its extremal values at the ends of the interval;
- write down a function which also has a local maximum in the interior of the interval;
- write down a function which also has a local minimum in the interior of the interval.

Now go back and make sure that, at each stage, the example you provide is not also an example at the following stage. For example, your first function cannot be continuous on $[0,1]$. (This stretches students' appreciation of counterexamples and the scope of a particular condition, as well as making them consider relationships among properties.) You may sometimes want to give your students impossible sequences, to see how they deal with them.

Notice that the task begins with a wide range of freedom of choice of example, then imposes increasing constraints (see *Theme: Freedom and Constraint*, p193). For more variants on this tactic, see Watson and Mason (1998).

The aim is to stimulate students to appreciate the subtleties of the concept through trying to construct their own examples. It also helps them to link ideas together by drawing on what is familiar in order to appreciate the unfamiliar.

Tactic: Expressing Generality

Tactic: Ask students what is common among several 'examples'.

For example, what is common among isometries of the plane and permutations of a set of objects? What is the same about, and what different about, certain exercises from a text?

Tactic: Ask students which aspects of a particular example of a mathematical concept are permitted to change and yet maintain its status as an example.

For example, $x \to |x|$ on the reals is not differentiable at exactly one point, and neither is $x \to |x+1|$; what could be changed, varied, or extended and still preserve this property? Ask them to express that generality either in their notes, to a neighbour, or even to the whole group (see *Tactic: Talking in Pairs*, p49). Furthermore, the function $x \to |x-1| + |x+1|$ is not differentiable at just two points. Can they extend this idea to a function that is not differentiable at exactly 3 points? At exactly 4? Can they generalise these examples to *n* points? Can they construct a function differentiable everywhere except at $\{1/n : n = 1,2,\ldots\}$, or at $\{n : n = 1,2,\ldots\}$? This type of example could then be extended to functions like $x \to \sqrt{|x^2 - 1|}$ and to 'glued functions' generally.

Another fruitful area involves developing examples of functions which are not continuous but which are Riemann integrable; for example, what is the difference between $\int_a^b \lfloor x \rfloor dx$ and $\int_a^b \lceil x \rceil dx$? Use these basic forms to construct more complex examples.

Over time students will begin to appreciate what they are being asked to do, especially if you then describe your thinking as you express a generality, stressing what can be changed and what is structural.

The idea is to take every opportunity to get students to express ideas generally, so that they become accustomed to it, and then to develop the habit of specialising again in order to make sense of a generality and reconstruct it for themselves.

See also *Tactics: Particular – Peculiar – General*, p88; *Invariance Amid Change*, p32; and *Themes: Mathematical Powers*, p184; *Invariance Amid Change*, p192.

Difficulty T3: Insufficient Facility or Competence	*Diagnoses*
'Students don't appear to spend enough time mastering techniques from the course. They don't seem to think of using them when appropriate, and they struggle to use them effectively.' 'Students don't seem to spend enough time working through their notes and making sense of the new ideas.'	There is more to mastering a technique than doing lots of examples. It is perfectly possible to follow a template but not really know what you are doing; it is also possible to understand concepts but not be able to choose specific techniques to solve problems involving them. (See *Issues: Developing Facility,* p180; *Doing is not the same as Construing,* p168; *Tactic: Advising Students How to Gain Mastery of a Technique,* p99.) Knowing whether to use a particular technique requires an appreciation of the contexts in which it is likely to appear, and what it achieves. (See *Tactic: Catching 'it',* p95; *Framework: Concept Images,* p190.)
Comment: An important aspect of learning mathematics is appreciating the scope of a particular theorem or technique. If the lecturer or text is always makes the transition from specific examples to the general case, then students may not realise that they too are being called upon to generalise. They may, for example, not realise that one reason they are expected to work through examples, and to construct examples for themselves, is so that they can realise (literally 'make real') the significance and the scope of the general statement (see *Tactic: Specialising, Generalising and Counter-examples,* p10). They may also not be clear about what must remain invariant for the theorem to apply, and what can change (see *Tactic: Invariance Amid Change,* p32).	

Tactic: Student Generated Exercises

Tactic: After you have been through some sample questions on a topic, ask students to make up (and do) their own questions of the same type. Variants include: asking for an easy and a hard question of that type, and the most general example the students can come up with; asking for a question which shows they know how to do a question of that type; and asking for a description of what constitutes 'that type'. You can tell from the features that they include in their questions and the features they leave out how good their grasp is of the general class of problems, and whether they are aware of the subtleties or difficulties which can crop up.

For example, a student who offers, in order to demonstrate their grasp of differentiation, 'differentiate $3x^2 + 2x + \sin x$' may not have got to grips with differentiating exponentials and logs, nor products, quotients and compositions of functions (see also *Tactic: Particular – Peculiar – General,* p88). This tactic applies to any type of question, whether a routine exercise or a typical 'theorem and application' from an exam paper.

The idea is to get students to shift their attention from reacting to whatever question is put in front of them and becoming absorbed in particularities, to examining the structure of the questions.

See also *Tactics: Boundary Examples,* p14, p136; *Assessing Degrees of Confidence,* p72; *Making the Most of a Worked Example or Exercise,* p79; *Advising Students How to Make the Most of a Worked Example,* p97; *Using Common Errors,* p10; *Advising Students How to Gain Mastery of a Technique,* p99; *Catching 'it',* p95; *Issues: Doing is Not the Same as Construing,* p168; *Developing Facility,* p180; and *Framework: Concept Images,* p190.

Difficulties with Concepts

Students have a variety of difficulties with different concepts, some of which have been studied in detail (for example Dubinsky and Harel, 1992, on functions; Leron, 1983, Solow, 1990, and Hannah, 1995, on proof; and Sierpinska, 1994). A structure for thinking about this more systematically is offered by the *Concept Images* framework (p190). This sub-section summarises several related kinds of obstacles students encounter. It starts with the use of words as technical terms, then moves on to mathematical concepts, compound concepts and processes becoming objects, before returning through the use of symbols and notation, to difficulties with definitions.

Difficulty C1: Technical Terms	*Diagnoses*
Students confuse *convex* and *concave*, especially when used without *upward* or *downward*.	The meanings of *convex* and *concave* depend on where you are 'standing'. Because their attention is drawn to enclosed or partly enclosed regions, most students will see a function like $y = x^2$ as concave. This is why some people use the additional word *upward* or *downward* to make this clear; $y = x^2$ is concave upward and convex downward.
Students find it difficult to distinguish between *independent* and *dependent, span* and *spanning set,* and *order of an element* and *order of a subgroup* . Students confuse *inverse* as reciprocal and *inverse* as inverse function.	If you are hurrying on to the 'important part', it is easy to overlook the value of emphasising the distinction between terms such as these. Even though the distinction may be clear to you, take care to make sure that it is clear to your students as well. Especially when someone is speaking quickly, there can be very little difference between 'independent' and 'in dependent', or 'span' and 'spanning set'. On the other hand, 'sets which span' and 'spanning sets' sound different but mean the same. In particular, students are often confused that an element can have an order when order is defined first for groups.
Some students carry over inappropriate associations with a term that has non-mathematical uses, while others ignore them even when they are relevant. For example, a *limit* is not necessarily a boundary or an extreme, while *basis* is not seen as a basis for expressing all other vectors.	Students do not have your advantage of frequent use and detailed understanding to help them decide when associations are relevant.

Tactic: Introducing Technical Terms

Tactic: Where a term draws on common usage, emphasise that usage when introducing it; where it does not, emphasise that it has a different mathematical meaning, perhaps by using an initial capital for a time until students are used to it.

Where there is potential confusion between two terms, try using only one of them until students are confident in manipulating it and using it to express themselves, then start using the other if appropriate (Baumslag, 2000, p176).

Appreciating the implications of a definition and what it admits or excludes takes time and experience. It is not something which can be passed over in a few moments as simply being the result of a formal statement.

Difficulty C2: What is a …?	*Diagnoses*
Students do not distinguish between a function, f, the value of a function at a point, $f(a)$, and the range of f. They expect a function to be specified in terms of a (simple) formula.	Students begin their study of mathematics by thinking about numbers. Functions, by contrast, require attention to processes or potential actions, which requires a shift in perception.
Students balk at specifying a function piecewise, as in $\left.\begin{array}{ll} f(x)=x, & x\geq 0 \\ f(x)=-x, & x<0 \end{array}\right\}$ or $\left.\begin{array}{ll} g(x)=e^{x}, & x\geq 0 \\ g(x)=e^{-x}, & x<0 \end{array}\right\}$. On seeing a sketch of this second function, students may also refer to it as a parabola because it 'looks like one' (see Dubinsky and Harel, 1992, p160).	Students were probably first introduced to functions through examples which used a single rule, possibly via tables, and sometimes in terms of graphs. Split definitions (glued functions) may strike them as complicated and they may be blocked by the complexity. There is a story that Cauchy rejected the first example as not being a function, until having it pointed out that this is the same as $x\sqrt{x^2}$ (which Weierstrass also thought of as $x\lvert x\rvert$). Euler did not accept such things as functions but did allow implicit functions (see Wanner and Hairer, 1996). However, one example does not always convince students, especially students with a strong commitment to a different perception.
Students do not recognise a function as made up of the composition of two or more functions (for example $\sqrt{1+x^2}$ or $\sin x^2$ – which is often confused with $\sin^2 x$ – and especially $f(g(x), h(y))$, and so are unable to differentiate or integrate complicated functions.	
Students find phase diagrams difficult to draw, interpret, or imagine.	Students are used to graphing position or perhaps speed against time, but not speed against position.
Students find changing their frame of reference difficult. For example, they find it hard to imagine what relative motion would look like if it was seen from a moving vehicle, rather than from a stationary position (despite experience of being in cars, buses, and trains).	Looking at something from a different perspective can be very challenging until you are used to it. Students need time to think through the mental imagery involved in making such shifts.

Comment: The concept of a function is just one of many that students stumble over. Linear independence and bases, limits, transforms, and matrices as linear transformations are just a few of the others. Students, like everyone else, stress some features (often based on past experience) and find it difficult to shift their attention to stressing other features. Meanwhile lecturers are often blissfully unaware that their students are not stressing or appreciating the same things as they are.

Tactic: Using Multiple Representations

Mathematicians use multiple representations of concepts for a reason: some representations are more useful for manipulating, others for imagining what is happening.

Tactic: Offer students several ways of thinking about an object or a situation. This can sometimes liberate a student who has a preference for one view over another. Mental flexibility is more important than finding the 'best' or 'right' way to think in any given situation.

For example, functions can be thought of as rules for calculating values, as tables of values, as graphs, as mapping diagrams, as a list of pairs, or even as pairs of lists in some cases. In combinatorics, a graph can be thought of as a diagram, a list of pairs, a matrix of vertices against vertices, or vertices against edges, or a family of sets. A group can be thought of as a collection of objects satisfying certain properties, a collection of actions upon a set of objects, or as an equivalence class of isomorphic structures. As a revision task, ask students to consider under which circumstances different ways of thinking are likely to be most useful.

Tactic: Say What You See

Tactic: Put up a diagram or a short worked example, and ask students to announce what they can see (without trying to be clever). The first time you do this it may help to give an example yourself of something you see (for example, 'there are three *x*s in the first line', or 'there are a pair of axes at right angles'). Sometimes it helps to use the rule that no-one may speak a second time until everyone else has spoken, so that it is best to 'get in quickly' while there are still easy things to point out. Anyone should be free to ask questions if they are at all uncertain about what is being described.

For example, you could use one or more diagrams which you think display the geometrical import of Rolle's theorem, the mean value theorem for differentiable functions, Cauchy's mean value theorem, or the Lipschitz condition. Hearing what strikes students may lead you to modify the diagram, perhaps by adding detail, or perhaps by breaking it down into a sequence of pictures or even an animation generated by Mathematica or Maple.

By hearing other students describe features that they themselves do not consider important, students have the opportunity to 'see things differently'. Sometimes students simply do not see certain parts of expressions, either because they are unconfident about them, or because they cannot cope with the implied operation (for example, square roots in the limits of an integral, or an integral in the denominator of an expression). Getting students to read out loud the statement of a problem or some complex expression can reveal conceptual difficulties when they pause or stumble over certain parts. Attention can then be directed to these aspects.

This is something students can use among themselves when studying. Further examples can be found on p30 and p91. When you are describing what you are seeing (either displayed or imagined), try using imperatives: 'imagine the graph of a function'; 'add a line parallel to …'; 'stop the point moving and …'. It is much easier to respond to the imperative than to other forms of verbs when working on images.

Difficulty C3: Turning Compound Objects and Processes into Single Objects	*Diagnoses*
Asked to prove something using cosets, students do not know where to start; some try to work with individual elements, others manipulate capital letters as if they were elements of the original group.	The notion of a set of functions, a set of group elements, or a set of vectors being itself an element in some other structure is sophisticated and requires a shift in the way you think. It is highly non-trivial.
Asked to consider solutions to a differential equation, students expect numbers rather than functions as solutions. Asked to work with a set of functions, students work with one or two members but not the set as a whole.	What comes to mind first is what you are confident about, not new ideas you have just met. Working with a complex entity like a function, a coset, a set of numbers, or a set of sequences requires a kind of 'letting go' of detail, and a stressing of the compound whole.
Asked to prove something in analysis, say about $\sqrt{2}$, students are inclined to use what they know from the past, rather than to work solely from formal definitions. Asked to interpret a graph, students think 'pointwise' rather than seeing x as varying and thus tracing out the graph, and they may see the point on the graph somehow as the function, rather than the height of the point as the function value. Asked to add two functions, or to think of a polynomial simultaneously as a function and as an arithmetic object (never mind a vector in a vector space), students have trouble coping with the different viewpoints.	Using a member of an equivalence class to name the whole equivalence class is also hard to keep a grip on, because it is easy to forget that the element you are manipulating represents a whole collection or class. There is a tendency to work with the name as if it were its old self, not the class it represents. Using a special notation such as $[a]$ or $\overline{a}$ can be of assistance, if only because it stimulates students to remind themselves what it means.
Being expected to think of linear transformations both as objects which can be added, multiplied, and scaled, and as transformations, students get confused as to when to think which way.	The notion of a linear transformation, or even a function, being both a process (a thing you do) and an object is sophisticated. It generally takes time to build up confidence in manipulating the process before it can also be seen as an object (see *Framework: Manipulating – Getting-a-sense-of – Articulating*, p187). Students are often blocked from seeing the wood for the trees by the presence of something which is not, for them, confidence-inspiring.

Comment: For more details on students' sense of graphs, see Nardi (2000). Sfard (1991, 1992, 1994) and Gray and Tall (1994) develop the notion of *reification*: the turning of a process or action into an object. For difficulties with *inf* and *sup* which affect students' perceptions of *limit* and hence their appreciation of analysis, see Nardi (2000a).

The idea is that, by identifying what they are stressing, you can find out whether students are stressing the right points. You can then choose to emphasise features they are overlooking, or to work on particular examples that indicate they are stressing inappropriately. By hearing what others are stressing, students get a chance to discover features of which they were not previously aware.

See also *Tactics: Specialising, Generalising and Counter-examples*, p10; *Boundary Examples*, p14, 136; *Say What You See (again)*, p30, 91; *Drawing Diagrams*, p92; *Catching 'it'*, p95; and *Framework: Concept Images*, p190.

See Tall (1992), and Nardi (1999).

Tactic: Inner Moves

Mathematical language serves a purpose, but students may not be aware of that purpose, nor of what to do as a result of it. It is therefore useful to be careful with statements such as 'Let ... be a ...'.

Tactic: When you notice a situation such as those just described, be explicit about the shifts that you perform automatically: 'I look at this polynomial as a function, then I think of each value as if it were a coordinate, ...', or, 'I have to multiply two sets of elements, which means that I have to multiply everybody in the first set by everybody in the second ...'.

Watch out for that succinct exhortation, 'Let ... be a ...', because it assumes that a lot of work has been done by the listener! When you catch yourself announcing to students, 'Let G be a group', 'Let r be a rational', 'Let V be a vector space', or 'Let H be a Hamiltonian', try to be aware of what you are thinking. You are probably prepared for the imposition of further conditions, ready to carry out allowable operations, expecting appropriate qualifications and thinking of a useful example on which to test out conjectures. Talk (briefly) to students about what you do when you hear a 'Let ... be a ...' type of statement. Many students apparently do nothing in response to such a statement, perhaps because they do not know what to do, or perhaps because they do not even know that there is anything to do.

Statements like 'Let ... be a ...' are only effective when used with terms with which the students are already reasonably familiar. When introducing concepts, it may be better to say something like, 'We need a vector space, which we shall call V', leading the students in more gently. (See also *Tactic: Introducing Definitions*, p24.)

Use a special notation to distinguish between a collection and its members, and be explicit about when you are calculating with a particular (typical) element, and when you are calculating with all elements in an equivalence class.

This is a special case of being aware of what you are stressing (and what, consequently, you are ignoring), and then being explicit to students about these (see *Tactic: Say What You See*, p20, 30, 91).

See also *Tactics: Introducing Symbols*, p90; *Advising Students How To Study a Mathematics Text*, p96; *Advising Students How to Learn How to Learn (Learning Files)*, p98; *Being Explicit about the Enterprise*, p56; *Theme: Shift of Attention*, p186; and *Framework: Manipulating – Getting-a-sense-of – Articulating*, p187. See Burn, Appleby and Maher (1998), p230.

Difficulty C4: Seeing Behind Symbols and Notation	*Diagnoses*
Students do not see $f(x,y)=2x+1$ as a function of two variables, and do not see $f(x)=3$ as a proper function.	The absence of a variable in the 'formula' can be upsetting at first.
Students tend to see $f(x)=(x-1)^2+2(x+1)-3$ and $g(x)=x^2$ as different and remain stuck with that first impression even after they have been shown algebraically that they are the same. With $(x^4-1)/(x^2-1)$ and x^2+1, students may not realise that the expressions are not the same, as the first is not well-defined at $x=1$.	Using different names for the same thing can be very confusing until you have become familiar with it. For example, 2/3 and 4/6 are different names for the same value and for the same ratio, and this carries over into equivalence classes and representatives. Similarly, $0.\dot{9}$ is another name for 1. The same (real valued) function can have many different expressions, but what matters is whether they give the same values for all elements in their domain. On the other hand, it is important that students do not gain a global perception that two expressions are the same without awareness of technicalities such as whether or not they are defined everywhere.
Students use quantifiers loosely or not at all, sometimes at the front, sometimes at the back, and sometimes both! For example, 'there exists an n such that $n(m+1)/2$ is even for all m' is ambiguous. Students do not distinguish between $(a+x)^2=a^2+2ax+x^2$ as an identity and $(a+x)^2=4a^2+ax+3x^2$ as an equation in either x or a. Students are perplexed as to when 'equating coefficients' is permissible (as in partial fraction decomposition) and when it is not, and make no connection with linear independence. Forcing two polynomials with parameters to be equal for all x is not seen as different from solving a polynomial expression for x.	Students do not appreciate the force or importance of implied quantifiers, nor the importance of their order. Students are not used to asking themselves, 'For what values of ... are these two expressions equal?'. This is a recurring theme, but one which needs explicit attention as it can be very off-putting for novices. Students may have been exposed to lecturers in different courses using quantifiers loosely. For example, 'for all n, $n(n+1)/2$ is an integer', and '$n(n+1)/2$ is an integer for all n' are equally valid and innocuous, but lead students into confusion through mixing them together.
Students are likely to overlook difficulties in meaning in an applied context due to units of measurement, such as $\log(x/y)=\log x-\log y$ when x and y are distances; what is the meaning of the log of a distance?	Students tend to leave units to the end and hope that they will all work out. However, users of mathematics are helped by paying attention to units, so if mathematicians skate over problems, students may be confused as to what they are allowed to use in a given situation.
Students are often confused about the difference between based vectors (representing a journey, say) and free vectors (such as forces).	This is a tricky distinction, requiring the conceptual leap of seeing that, although forces act at a particular position, the vectors that represent them can be moved so as to add them together.

Switching rules from one context to another can also be problematic; for example, students may get the idea that anything which holds true for real numbers can be naturally extended to complex numbers. While this is often the case, applying $\sqrt{a}\sqrt{b}=\sqrt{ab}$ to a complex-valued expression can lead to notational paradoxes such as the following.

$$\sqrt{-1}\times\sqrt{-1}=\sqrt{\frac{-1}{1}}\times\sqrt{\frac{1}{-1}}=\sqrt{\frac{-1\times 1}{1\times -1}}=\sqrt{\frac{-1}{-1}}=\sqrt{1}=1$$

Tactic: Reconstructing and Re-expressing

Tactic: Try getting students to express in symbols something written in words, or to reconstruct a commentary from just symbols. Try getting students to draw a diagram from a verbal description, or to describe for someone else how to draw a diagram that only they can see.

The idea is to stimulate students into converting between words, symbols, and diagrams, and to get them to expand technical terms and symbols into words with which they are comfortable and which they find meaningful but which are also mathematically correct. In the process they may come to recognise that the same thing can be expressed in several very different forms. Also, they may learn to read diagrams rather than simply to look at them (see *Tactic: Say What You See*, p20, 30, 91).

See *Tactics: Expressing Generality*, p14, *Invoking Mental Imagery*, p55; and *Framework: Manipulating – Getting-a-sense-of – Articulating*, p187.

Difficulty C5: Definitions	*Diagnoses*
Students have difficulty remembering the definition of the order of an element in a group and the order of a subgroup of a group.	Students have not internalised the idea of the order of a set as the cardinality of the set, nor have they made the link between the order of an element and the cardinality of the set it generates. Very often these notions go by in quick succession, with no time to really work on the connection.
Students do not think of a polygon as being allowed to be concave, or to be self crossing, or to have vertices with an angle of 180°. Similarly, they have trouble accepting the idea of a triangle as a special case of a trapezium, a magnetic monopole as a dipole with one end infinitely far away, a particle with zero mass, or a Dirac delta function.	Students have strong ideas of what is 'sensible', and do not always appreciate that, sometimes, considering something as an 'unphysical' limit of something else can be helpful. It is perhaps worth remembering that there are examples of this even in nature: glass may appear solid, but is in fact a very viscous liquid, and will flow if given enough time.
Despite the formal definition of a function, students treat functions as being specified by equations.	When a definition (such as function) incorporates but extends familiar examples, it is natural not to pay much attention to what else has been included. There are many examples of this in mathematics. The notion of 'pathological examples' even enshrines this notion.

Difficulty C5: Definitions (continued)	*Diagnoses*
Students have difficulty recalling all the details of definitions, especially where they are a little complicated or when there are several equivalent versions, as in the definition of a limit, an integral, or a subgroup.	Appreciating the role of conditions or restrictions and internalising complex definitions requires work on the reasons for including and excluding different possible features.
Comment: Students often think that definitions arrive on tablets of stone. They do not appreciate that a good definition is one that permits theorems to be stated succinctly (Lakatos, 1976). They do not think of 'degenerate cases' as being acceptable. They tend to approach definitions as referring to examples with which they are familiar, without looking to see what else has been included.	

Tactic: Introducing Definitions

The classic structure of mathematical exposition consists of a definition, then a lemma or two, and then one or more theorems. This may work for experts, but for students it sometimes makes the subject seem dry and inflexible, with little room for creativity. However, the formulation of definitions is a highly creative aspect of mathematical thinking.

Tactic: Try offering some examples and some non-examples of a concept, and then get students to identify the features which are required, and so to participate in formulating a definition.

Try starting with a theorem, stated in general terms. Point out where it is necessary to gain precision so that there is a chance of actually proving the theorem, and offer some examples for which the theorem works. Invite students to consider what it is about the examples which makes the theorem work, and so to formulate a definition for themselves from those examples. Then compare their definitions with the one you are going to use, and point out the role that any of the extra features in your definition are likely to play in the proof. During the proof, draw attention to where those features are needed.

For example, in proving that there are only five regular polyhedra in three dimensions, it is necessary to be very clear about what the term *regular* means. Many students think that a regular polygon simply has to have all the edges equal, or even that it must simply have some symmetry.

In defining arithmetic, geometric and harmonic means, it is helpful to recognise that each is trying to use a single number to capture some feature of a set of numbers, and that each of these means is a conjugate of the others (the reciprocal of the arithmetic mean of the reciprocals is the harmonic mean, and the exponential of the arithmetic mean of the logarithms is the geometric mean).

In defining a random variable, it is necessary to appreciate that it is actually a function. In defining a probability density function it is essential to realise that for an infinite sample space, some approach is needed which does not try to specify a finite probability at each point of an infinite set.

Note that working formally from definitions, through lemmas, to theorems and their proofs does not constitute *being explicit about the enterprise* (see p56). To be explicit about the formal enterprise requires cross-checking between intuition and the consequences of formalisation. Thus the formalisation of a sense of *function* admits glued functions, functions specified as the limit of a sequence of operations (as in the von Koch snowflake curve or a blancmange function), wildly discontinuous functions, and so on. Further classifications via definitions (continuous, integrable, bounded variation, or measurable) are used to outlaw objects which the original definition permits. If students only ever encounter nicely behaved functions they can draw on their calculators, they are unlikely to appreciate the need for all the rigour and formality of calculus.

Tactic: Intensive and Extensive Definitions

An intensive definition is a description that captures the essence of an idea. For example, seeing continuity of a function as meaning that it has no breaks or wild oscillations; seeing a *basis* as a vocabulary in terms of which every other vector can be expressed without redundancy. To check that something meets an intensive definition requires knowledge and intuition about the object.

An extensive definition is stated formally. To check that something meets an extensive definition only requires understanding of the meaning of the individual words comprising it and some calculational competence. A formal definition of continuity is extensive, since checking it is essentially a clerical exercise: given an ε, you have to show how to find a δ, and similarly for other equivalent definitions. (See also *Tactic: Introducing Symbols*, p90.)

Tactic: Try offering students an intensive definition, with some examples, perhaps presented on one OHP or board, and then an extensive version on another OHP or board. Indicate the shift of perception required in order to formalise the informal.

For example, in the case of continuity, it is usual to shift from a dynamic sense of a function to focusing on a single point, and then to capture the meaning of 'arbitrarily close'.

Tactic: Using 'Strange' Examples

The title of this tactic uses *strange* rather than the more common *pathological* in order to take away from the impression that such examples are unusual or have something wrong with them.

There is considerable controversy between lecturers as to whether it is advisable to show students strange examples in calculus, but the same question applies to every mathematical topic. On the one hand strange examples are usually complicated and so students cannot see why they should make an effort to understand them, and, in any case, most students are unlikely to run into these sorts of examples in other disciplines. On the other hand, if students do not appreciate the range of possibilities encompassed by a definition, they cannot appreciate the import of the theorems, or the technical difficulties involved in proving those theorems.

Tactic: Offer students unfamiliar examples, pointing out which aspects of them are useful for disturbing one's cosy sense of, for example, what a function is, and clearly state the conjectures to which they provide counter-examples. Find a way to be interested in the examples yourself, perhaps by focusing on the generalisable features of the particular examples. Construction of examples highlights the constructive aspects of otherwise technical definitions.

For example, in group theory, direct products, and later semi-direct products, are essential for appreciating why some theorems are hard to prove. In linear algebra, constructing vector subspaces that are not in 'standard' positions helps students appreciate the generality of the theorems.

For example, the simplified square wave functions

$$x \to \left\{ \begin{matrix} 1 & \text{for} & x \in [2n, 2n+1] \\ -1 & \text{for} & x \in [2n-1, 2n] \end{matrix} \quad n = 0, 1, \ldots \right\} \text{ and}$$

$$x \to \left\{ \begin{matrix} x - 2n & \text{for} & x \in [2n, 2n+1] \\ 2n - x & \text{for} & x \in [2n-1, 2n] \end{matrix} \quad n = 0, 1, \ldots \right\}$$

are harder to express in symbols than to display as graphs. Students could be asked to extend other functions (specified only on one interval) to periodic functions on the reals, in order to get to grips with these types of function specifications.

Functions such as $x \to \left\{ \begin{matrix} x \left\lfloor \frac{1}{x} \right\rfloor & x \neq 0 \\ 1 & x = 0 \end{matrix} \right\}$ and $x \to \left\{ \begin{matrix} x^2 \left\lfloor \frac{1}{x^2} \right\rfloor & x \neq 0 \\ 1 & x = 0 \end{matrix} \right\}$ tend to have interesting properties at and near $x = 0$. Some of them are differentiable at 0, with slope 0, yet arbitrarily close to 0 there are points at which they are also differentiable, but with arbitrarily large slope! Another function with this property is $x^2 \sin(1/x)$. These examples challenge an unspoken intuition that at points close to a point of differentiability, the slope will not vary significantly.

Tactic: Reconstructing Definitions

It is likely that students' definitions will reflect the range of examples with which they are familiar, but this may allow more or be more restrictive when compared to the conventional definition. For example, students' recall of the definition of differentiability is unlikely to include functions which arise as the integral of a discontinuous function.

Tactic:	Ask students to write down a definition, in their own words, of some specified term that has been used (and defined) previously. Then get them to exchange definitions with a fellow student, and to find an example which shows that the definition as stated is either too restrictive or too inclusive. In an introductory course this can be done very effectively with an ordinary concept such as a number, an equation, a polygon, or a statistical distribution, in order to awaken students to the need for precise definitions in mathematics.

Difficulties with Logic

Difficulty L1: If and only if; necessary and sufficient; for all there exists vs. there exists for all	*Diagnoses*
Students do not recognise that *iff* requires two things to be proved. Students confuse *if* with *only if* and vice versa. Students check one but not both conditions. Students do not understand *necessary and sufficient* conditions, and do not relate this to *iff.*	When you find that something is difficult or has little meaning, it is natural to try to find a mechanical way of dealing with it. Students find it difficult to accept that *iff* implies two steps: *if* and *only if*, so they just jump at an interpretation for *iff.*
Students do not distinguish different meanings for quantifiers in different orders.	Sometimes experts in a hurry forget about quantifiers, and then put them at the end of the statement rather than at the beginning. This suggests that they are attending to and expressing the meaning of the whole statement, not attending to the components of the statement. However, the person they are speaking to has to attend to the components in order to work out the global meaning.
Comment: Attempts to teach logic first and content second have rarely succeeded. Logic is something you have to learn in context. However, logical reasoning is difficult for many students, so explicit attention needs to be given to it every so often (see *Tactic: Directed – Prompted – Spontaneous*, p91). Some people recommend never using *necessary and sufficient* on the grounds that most students and some lecturers find it confusing.	

Tactic: Distinguishing Common and Technical Meanings

In English, the words *a*, *any*, and *all*, can be very confusing. 'Let *x* be *any* number' can be interpreted as referring to some particular but unspecified number. Mathematics uses the added sense of indifference as to which number, hence the use of *any* to indicate generality, something which applies to *all* numbers. 'The derivative of *a* function need not itself be differentiable' can be interpreted as 'there is one such function', and also as 'in general', but compare this with 'the sum of the angles of *a* triangle is 180°', in which the *a* means *all.*

Tactic: Try to be aware of the words you use to signal generality, and check that students are not misinterpreting you. Discuss the difficulty explicitly once or twice when you notice yourself using *any* and *a* frequently.

Many mathematical terms, such as linear, set, group, field, ring, continuous, function, or differentiable, have common meanings in English. Sometimes there are links between the meanings, and sometimes there are significant differences.

Tactic: When introducing a new term to students, take a brief moment to indicate its origins, or the sense (if any) in which the ordinary usage is being made more precise. Where the usage is different, indicate what image or sense the word is trying to capture.

For example, the *socle* of a ring is a foundation or pedestal on which the structure can be said to stand, the term coming from the French for plinth; the use of Z for integers comes from the German word *Zahlen,* meaning counting; what is 'open' about an open set, and what images can usefully be associated with the word? On the whole it is not useful to draw students' attention to other meanings of the term which are not employed in the mathematical usage, as they may inadvertently find this other meaning sticks and then gets in the way in the future.

Tactic: Translating Technical Terms

The *if and only if* construction takes considerable getting used to, because of the oddness of *only if.* When the double arrow $A \Leftrightarrow B$ is used as well, students often have trouble deciding which arrow corresponds to *if* and which to *only if.* If *necessary and sufficient* is used as well, this only compounds the confusion. One approach is to stick with the implication arrows, perhaps in addition to a Venn diagram. Another approach is to stress the explicit use of 'what do I know' (from the problem) as the starting point, from which to deduce 'what do I want'. Translating an emotion-free topic into more emotionally laden terms can help students 'feel' the force of the logical terms. This may enable them to see both the role that *if* plays (tells what additional things they know), and enables them to deal with it. *Only if* is more difficult still; students need help in finding familiar situations in which A is true only if B is true. Again, the language of *know* and *want* can be of assistance. It can also help students remember to check arguments in both directions.

Tactic: Use technical and informal equivalents side-by-side, then gradually reduce the translation into informal terms until your students are able to cope with the technical terms alone.

It is vital that students gain facility with the terms mathematicians use, so a process of starting with informal and explicit remarks, then gradually increasing the degree of formality is needed so that students are weaned off explicit support (which they can invoke for themselves if needed) until they find themselves spontaneously using the technical terms. (See *Tactic: Directed – Prompted – Spontaneous,* p91.)

Difficulties with Studying

Students who have succeeded in school through native wit and mathematical ability may not really appreciate how to go about studying mathematics when it starts getting difficult. Here are some of the main symptoms.

Difficulty S1: 'We need more examples, more exercises.'	*Diagnoses*
Students say that they need more examples for a particular topic. Often this is because they cannot find an example which closely matches the assessment questions.	When a student is shown an example, they may or may not see it as exemplary. This is because they may not yet appreciate the general concept which is being illustrated, and are caught up in the details instead. Do they recognise what is allowed to change and what must remain the same in your example? Are they paying attention to the same parts as you are (perhaps subconsciously), and have you been explicit about what you are mentally stressing?
Comment: For you, an example is an example *of* something, usually a general type, principle, or object. However, unlike you, students may not yet appreciate the generality; for them, the example(s) are a major route to appreciating the generality, but their attention may be focused on the details. They may not see through the particular example to the general. Beware also that a beginner may notice and focus on aspects that an expert knows are of little or no importance.	

Tactic: Making an Example be Exemplary

Tactic: When working through a typical exercise or problem, take care to make clear to students the choices you make at each stage. What is it that makes you think of taking the next step? This is often what students cannot get from texts or from rapid-fire lectures. Take a moment to indicate what it is that makes the example 'typical', and what features are particular to the example itself.

Tactic: When describing an example of a mathematical object (such as a group, vector space, metric space, or space of functions), take time to stress what it is about the example which makes it exemplary. How could it change and yet remain an example, and what features cannot be changed if it is to remain an example?

For example, the collection of permutations of a finite set under composition is a group, but the size of the set does not matter, and the contents of the set do not matter. If some of the elements of the set are painted specific colours, then the permutations preserving colour yield a sub-group of the original. Similarly, the set of polynomials of degree at most n with rational coefficients form a vector space over the rationals but not over the reals. However, allowing the coefficients to be real does produce a vector space over the reals. Students are often perturbed because polynomials can, in their minds, be multiplied together, but this is not used in the vector space.

Tactic:	When working through the steps of a proof on an example, indicate at each step what it is about the example which allows the proof to work, and what is specific to that example. Try shifting the responsibility for this onto students (perhaps in a tutorial).

- See *Tactics: Specialising, Generalising and Counter-examples,* p10; *Expressing Generality,* p14; *Student Generated Examples,* p16; *Advising Students How to Make the Most of a Worked Example,* p97; *Particular – Peculiar – General,* p88.
- See *Issues: What is Exemplary About an Example?,* p173; *Discussing Mathematics,* p176; *Tactic: Say What You See,* p20, 30, 91; and *Theme: Invariance Amid Change,* p192.

Difficulty S2: Drawing and Reading Diagrams	*Diagnoses*
Students do not find your diagrams particularly helpful; students do not, or cannot, draw diagrams for themselves; students do not seem able to make use of diagrams to read off information. Students find it difficult to read and interpret speed-time, position-time, and speed-position graphs, or even to distinguish between them.	When students look at a diagram drawn by someone else, they see only the finished product. They have to learn which features are significant and which irrelevant, what is allowed to move or change, and what relationships must be maintained. Diagrams are not just pictures; they have an implicit structure, connected with the order in which the elements are drawn, the relationships between the components, and the sorts of things one thinks about as one draws those elements.
Students do not appreciate the difference between a particular diagram offering a pertinent example and a diagram showing a generic example.	Most diagrams are generic in the sense that they illustrate something that is invariant, even though some other things are permitted to change. It is vital, as with any example, that students appreciate what can change and what must remain invariant if they are to make sense of the diagram. A good diagram may be worth a thousand words, but only if the student appreciates its content, and has perhaps tried to articulate it in words for themselves.
Students see diagrams *as* proof.	It is a subtle matter to distinguish between a diagram which helps to organise one's thoughts, even being able to be read as a proof, and diagrams *as* proof.

Tactic: Say What You See (again)

Tactic:	Get students to say what they see in general, but also to say what they think the meaning is of particular points in relation to the topic that is being illustrated.

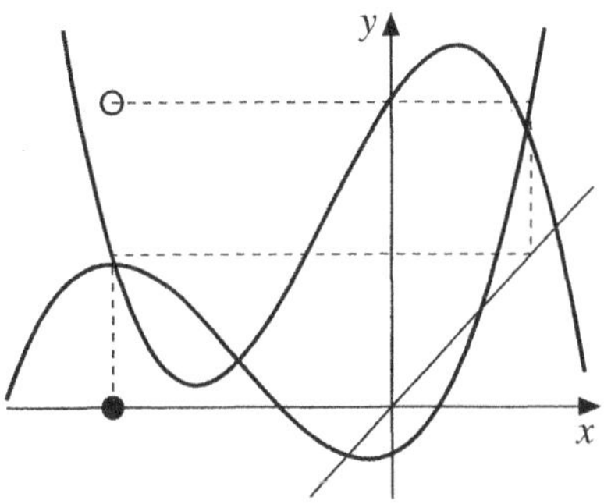

For example, invite students to imagine that the solid dot on the x-axis at the end of the vertical dotted line moves back and forth along the axis and that the rest of the construction 'follows' it while the functions remain fixed. Ask students to describe to each other in pairs what features they can see, then get them to announce these to the whole class. (It is amazing what is and is not noticed!). Then get them to think about what happens to the hollow dot lying above the initial dot on the x-axis as the solid dot moves along the x-axis and the construction follows.

By tracking coordinates, they can find the function whose graph is traced out by the end point. Then they can try to draw similar pictures for other functions (for example $f^{-1}gf$, or $f(gh)$).

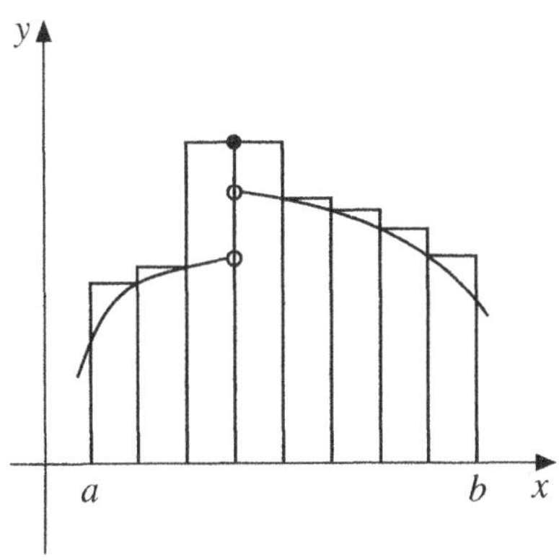

As another example, invite students to say what they see in this diagram, first to each other, then to the whole class. Then get them to imagine the interval widths getting smaller and smaller, but always with one interval end point at the discontinuity. Pause while you and they do this mentally. Again, get them to describe to each other, and then possibly to the whole class, the effect of the single discontinuity on the outer sum.

Then get them to superimpose the lower sums for the same interval on top of the picture, and again to imagine the intervals getting smaller and smaller. What would happen if the intervals never had an end point at the discontinuity? What sort of a function would *not* be Riemann integrable then?

Getting students to say what they are seeing often reveals what they think is important, but it needs to be done in an atmosphere of acceptance of any detail, not of people trying to show off to each other. Sometimes someone may need to ask for clarification of a detail, and so discussion can be generated.

Following a trail of details (such as tracking coordinates in the first example, or imagining the intervals getting smaller and smaller in the second), working out a sequence in which to draw the diagram (through tracking interdependencies of details), and constructing an overall story for what the diagram actually illustrates are all essential parts of making sense of the diagram, which students may not do naturally.

Say What You See can be used on any complex algebraic expression or diagram (see *Making Links*, p159).

Tactic: Invariance Amid Change

Tactic:	Ask students what features of a diagram can change and while still allowing it to retain a particular property. The same applies to any example of a theorem or technique.

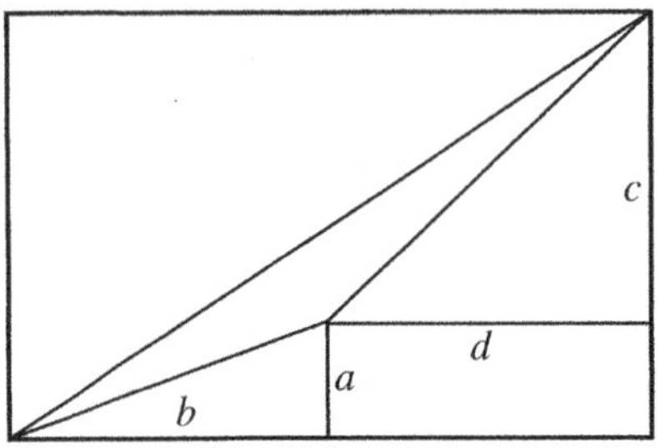

For example, in the diagram, the slopes of the two segments from corners of the outer rectangle to the inner one can be compared to the slope of the diagonal of the rectangle. How could this be expressed in words? Where can the inner point move to and still preserve the relationship between the three slopes?

It is also worth remembering that a number of conventions in mathematics are due to a desire for invariance, and the simplicity and elegance that it brings. For example, mathematicians could use a number other than e as the basis for exponentials, but e^x is invariant under differentiation and integration, whereas there is some $m \neq 1$ for which $\frac{d\left(a^x\right)}{dx} = ma^x$ for any other number a. Similarly, the use of radians to measure angles makes $\sin x$ easier to manipulate, as it then has slope 1 at the origin.

See *Tactics: Specialising, Generalising, and Counter-examples*, p10; *Invoking Mental Imagery*, p55. See *Theme: Invariance Amid Change*, p192.

Students can often find themselves caught up in a treadmill of other people's explanations and exposition, of doing exercises and receiving marks. If they are not encouraged to use their own powers of mathematical thinking, to construct examples, to express generality, and to particularise, they are likely to become more and more dependent on template examples and clear exposition. Teaching and learning can be seen as two separate enterprises, or as a joint enterprise in which the teacher or tutor's job is to progressively 'get out of the way' of the student, providing input and stimulation, tasks and feedback which reward the student for using their own powers.

Difficulty S3: Not Remembering	*Diagnoses*
Students claim not to have seen a topic that you know they must have encountered. Students fail to recall a topic or definition that they have used reasonably competently previously. Students remember parts of a theorem but forget others, especially technical conditions.	Most students tend to make sense of things using what is nearby, without drawing on past knowledge. In order to get them to recall something from the past you have to make some sort of link or trigger which gives you access to that memory. One way to create such triggers is through intentionally forging links as you go along. Doing a large number of exercises may or may not result in consolidation or the requisite links.
Comment: While it is tempting to offer students 'hints', 'suggestions' or 'guidance', it is first of all important to decide what your attitude is, because your language and gestures will convey a message. Is mathematics all worked out, meaning that students are to be given hints as to what you have in mind, or are you offering guidance? In the latter case, how do you know what to suggest at that point? Students need to know what makes you think of a suggestion or idea, not just the particular idea; otherwise you will end up training students to be dependent upon your hints or suggestions.	

Tactic: Forging Links

Tactic: After working on a topic or a problem in which some useful technique or approach has emerged, suggest to your students that they pause and imagine themselves working on another problem and having that technique come to mind. Ask them to imagine themselves using that technique again. The technique could be mathematical or could be a sense-making technique such as looking for invariance amid change.

This will be more effective if you have discussed, or they have thought about, the conditions under which the technique or approach is effective.

Difficulties with Non-routine Problems

Students who can do routine problems perfectly are not necessarily prepared to tackle non-routine problems which require juxtaposing ideas, adapting familiar techniques, or are simply more complex than the ones with which they are familiar. Avital and Barbeau (1991) give examples of students approaching problems inappropriately and they suggest reasons for these that could be worked on with students explicitly. Some institutions have introduced a course designed to develop mathematical thinking, to expose students to the historical, cultural and philosophical issues in which mathematics is embedded, and even to raise pedagogical questions, rather than to introduce new mathematical topics. Such courses are based around problem solving, and they have proved to be very popular (Yusof and Tall, 1994). Students reported that, for the first time, they found themselves actually doing new (for them) mathematics, actually being mathematicians rather than merely absorbers of worked-out techniques. When such a course is presented in their third year, students often say they wish they had had the course earlier in their programme.

It is certainly not necessary to confine problem solving or historical, cultural, philosophical and pedagogical issues to a single course. Indeed if problem solving is seen as central to mathematics, then it is vital that this perspective imbue every course. Providing a course focused on problem solving, history, or pedagogy can help those students who might be attracted into teaching to get a taste of the active aspects of mathematics. Students intending to pursue mathematics could also benefit from developing their problem solving skills, while students using mathematics in other disciplines could benefit from using mathematical thinking to tackle problems based in their discipline. This might also help them to appreciate more fully both the mathematical techniques they are being taught and, more importantly, the role of mathematics in their discipline.

Halmos, as ever the provocative and thoughtful mathematician and teacher, suggested that

'A teacher who is not always thinking about solving problems – ones he does not know the answer to – is psychologically simply not prepared to teach problem solving to his students.'

(Halmos, 1985, p322).

I would go further and suggest that they are probably not prepared to teach any aspect of mathematics effectively.

Difficulties with Applications

Students may have been exposed to a variety of techniques, and displayed competence in their use, yet not have them come to mind when asked to apply them to new contexts.

Difficulty A1: Locating Forces	*Diagnoses*
Many students have considerable difficulty in working out where forces are acting and in which direction. For example, the notion that, when you lean against a wall, there is a force exerted on you by the wall can be very hard to accept.	Newton's laws are far from obvious, yet once accepted they seem so natural that it is easy to overlook the struggle that some students have in coming to terms with them.
Many students do not have a physical sense of how pulleys and other physical objects work. They may not have learned to recognise the presence of forces in everyday actions, or to appreciate the differences between objects made from different materials (for example between plastic and glass, or wrought iron and steel).	Awareness of the properties of physical objects may only develop when attention is drawn specifically to them, and when students attempt to use objects in a laboratory setting to achieve a specified aim. Many students have difficulty visualising forces and fields. Also, adding forces requires them to turn based force vectors into a free ones, and this can be a difficult conceptual leap.
Comment: It is one thing to recognise forces in laboratory or classroom situation, but quite another to recognise them outside the classroom. Pre-constructed experiments in a laboratory are not always successful in providing students with experience of how forces are identified, so some students may need continued encouragement and assistance in visualising what is happening in a given situation. Having a 'feel' for the situation is often vital in being able to express it in terms of forces and then to produce a useful mathematical model.	

Tactic: Forcing Forces

Tactic: In the last five minutes of each lecture for several sessions, display a common object (a can opener, corkscrew, retractable pen, or bicycle, for example) to your students. Invite them to imagine what is happening in a part that is obscured or hidden or to consider what forces are acting, and then talk about how you think about all the forces. Also, try giving your students the force diagrams for some simple situations and see if they can work out different contexts or settings that would give rise to those same diagrams. A similar process could also be used in other situations, for example: carrying out some actions and challenging students to specify a group from which the actions are taken; showing a video of queues building up and challenging students to specify a relevant distribution; showing the graph of a function and challenging students to identify its formula.

It is necessary to appreciate the physics, chemistry, or biology of the problem situation in order to select a suitable mathematical answer. For example: 'the function $V(x, y)$ represents the potential energy of the system and must therefore be bounded', 'the potential must be symmetric about the axis and flat at infinity', or 'we reject this solution because it is physically meaningless'. However, the last statement can be dangerous: many concepts that at one time were considered physically meaningless (complex roots to equations, negative energy) have later acquired meaning in certain situations. For example, complex roots can signify a quantity with both a magnitude and a phase, and negative energy electrons can be identified as positrons.

Such reasoning requires a sophisticated understanding of physical principles, but it also requires an equally sophisticated insight into the way that these principles affect the mathematics in the model.

Many students do not 'see' forces acting in the world, and many have great difficulty with imagining physical situations. However, repeated exposure to the fact that it is a valuable skill, and to how an expert does it, can influence students' way of seeing the world. Even if students have a 'feel' for the way certain objects behave, such as a wheel rolling and slipping, or pulleys and levers, this sort of intuition is difficult to capture in mathematical symbols when trying to formalise the analysis of a problem. Not having such intuition renders much of mathematical analysis meaningless for students. For example, the awareness that many physical properties of matter are independent of sample size is essential, and so obvious as to be not consciously thought about by most lecturers, yet many students do not appreciate it.

Tactic: Developing 'Physicacy'

Tactic: Look out for films and computer animations of large-scale phenomena, such as beams bending and twisting, bridges and aeroplane wings vibrating, and wires stretching, and arrange for students to see and discuss these.

Tactic: Before drawing or showing your students a picture of a model of a situation, ask them to imagine it for themselves. Then take time to be explicit about how you visualised, imagined, or constructed the model as you built up your picture of it.

For example, the vector field around two bar magnets in three dimensions might be built up by imagining them separately and then slowly bringing them mentally into range, or by working out what forces would be experienced by an object in the field; the magnetic field around a wire carrying a varying electric current could be built up by considering the static case of a constant current, and then allowing that to vary.

Difficulty A2: Approximating and Checking	*Diagnoses*
Students find it difficult to estimate or approximate in order to check the reasonableness of an answer.	Students are not experienced in mentally simplifying and estimating. They are simply committed to finding the answer.
Students display reluctance to check their work, or to approach a problem from two different directions.	Students who have struggled to get an answer may not want to check it because, if it is wrong, then they may feel obliged to do more work; it seems easier to hand it in and later find out what they should have done.
Students are misled about the shape of a function or about the convergence of a sequence by the limitations of their calculator screen.	Students need to discover that calculator and computer screens are only ever approximations when displaying graphs and decimals.
Comment: Students may need to be in the presence of a tutor who identifies and comments on the processes they go through mentally in order to reach an estimate, and who takes the opportunity to approach problems from two directions in order to get a check on the reasonableness of their answer.	

Tactic: Estimating Explicitly

Tactic: Set your students an estimation problem for homework at the end of each lecture for several sessions, and have your answer on the board when they arrive for the next lecture. Spend a moment being explicit about the kinds of approximations and estimates that you have made. Try to ensure that you have made estimates in two different ways, as a check of reasonableness.

For example, Enrico Fermi made a collection of estimation tasks for honing his students' skills, such as the number of piano tuners in a city or town of a given size; the number of bakeries; the number of dry cleaners; the number of bricks in a particular large building; the number of grains of salt in a given salt cellar; or the number of people between 50 and 60 in the country.

Difficulty A3: Modelling	*Diagnoses*
Students find it difficult to move from a problematic situation to a mathematical model, for example in solving a problem involving pulleys, or sliding and rolling. They don't seem to know where to start; they are not confident about where forces act or in which direction.	Students may not have a well-developed sense of the particular situation, and how the forces work. As tutor, you probably already know what mathematical model is most appropriate, whereas students have still to see through the particular aspects of the problem and locate the mathematical essence. To do this they need to form some sort of a mental image of the root of the problem, which means selecting pertinent variables, removing complexity, and so on. Then they need to express that image mathematically. Modelling requires movement back and forth between the world of the original problem, mental images of an idealised situation, and the mathematical model.
Comment: Re-discovering what it is like to be a student in the face of a problem is non-trivial. It is most easily done by placing yourself in a similar position: take a problem that you have not worked on before and pay attention to what you do, how you think, what you try, what you reject and why.	

Tactic: Modelling Explicitly

There are many situations in applied mathematics in which some degree of modelling, of re-casting a problem in a form and language to which known techniques can be applied, is necessary in order to reach a solution. While the author or lecturer may be aware of what is possible technically, students may not. Consequently, some of the simplifying assumptions may be quite mysterious. Often each step towards simplicity is actually a contribution to complexity from the students' perspective.

It is worthwhile drawing simplifying assumptions to students' attention, as many may be mystified as to how the final, tractable model is arrived at, starting from the original problem. It may not be at all obvious to a student why the chosen assumptions are the right ones to make in a particular circumstance, and this is further complicated by the fact that the teacher may regard certain assumptions as so obvious as to be not worth mentioning.

Tactic: Look for opportunities to highlight the assumptions and simplifications that have been made. Talk to your students about how you know to make those simplifications or other assumptions. Be prepared to explore the consequences of your students' proposed alternative assumptions, at least to the point of seeing why the mathematical analysis would become too difficult. It is can be very difficult for students to appreciate that choices in modelling are often based on ensuring that the technical skills of the modeller can be used.

For example, take the case of a ball thrown horizontally. Certain features come immediately into a tutor's mind: Newton's laws, uniform acceleration, particles, lack of air resistance and so on; but how many of these are also in their students' minds? Indeed, these may well be inappropriate initial assumptions if we add that the ball is small and made of polystyrene, that the ball represents a particle in an accelerator, or that it is thrown by an astronaut in space.

As another example, take the case of transverse vibrations in a string. First, we might totally ignore the effects of gravity (but not mention that this is what we have done), then assume that we can model a short section of string as a particle, and finally assume that the particle moves only in a direction perpendicular to the line of the string. These are justifiable assumptions, but, if they happen too quickly, students may be left with little except a feeling of unease and personal inadequacy.

Reflection

Task: Reflection

Try to recall the essence of two or more tactics, and, for each, imagine yourself as vividly as possible using one during a session.

Comment: Imagining yourself as vividly as possible in a typical situation using a tactic you want to try out is the best way to open up the possibility of actually thinking of using it when it matters.

Be explicit to your students about your use of the tactic, so that they become aware of the possibility of using it themselves.

It might appear that use of these tactics will eat up valuable class time. However, if students are struggling with basic ideas, they are going to find the ideas that depend on them even more mysterious. Consequently, it is more efficient to spend time establishing not only awareness and sensitivity, but also ways of working, so that students can take more responsibility for their own learning as the course develops. See *Tactic: Directed – Prompted – Spontaneous*, p91.

2 Lecturing

Introduction

In a survey of third year mathematics undergraduates on questions appropriate to first years, Johnston Anderson and colleagues came to the following conclusion.

'The most striking aspects of this experiment were that, on what most of those teaching first-year undergraduates would consider to be essentially simple questions, only about 20% of the responses were substantially correct and almost 50% did not contain anything that could be deemed to be minimally "credit-worthy".'

(Anderson et al., *1998)*

Is this true of your institution? Is it worth finding out? Have you ever asked students in their second or third year whether they recalled a topic from a previous year, and found that many denied having met it, even though the lecturer affirmed that it had been covered? Anderson *et al.* drew some conclusions.

'This suggests that a considerable amount of what is taught to mathematics students in general, as "core material" in the first year, is poorly understood or badly remembered. It is as if the experience of students, attending one module after another, is such that they tend to "memory-dump" what they have had in previous modules, rather than retain it and build it into a coherent knowledge structure. In many instances, there was very little that the students actually could recall, even in cases where they had gone on to do further study in the same topic later.'

(Anderson et al., *1998)*

So what are lectures for? Are they the principal means of transmission of ideas from lecturer to student? Are lectures an opportunity to inspire students by raising problematic issues and by demonstrating the role and significance of major results and techniques? Are they an opportunity to work in depth on some major topics, with the assumption that students will then work in a similar way on other less central topics? Are they structured so as to be as effective as possible?

What do students think lectures are for? Are they carrying forward their experiences from school, expecting to understand everything in class, or are they beginning to learn how to work on mathematical ideas for themselves, inspired and informed by texts and lectures?

Is it best to hand out notes? If so, how complete or brief, and when is it best to do it?

The suggestions made in this and other sections are intended to support you in stimulating students to really understand what they are doing so that they can recall or reconstruct most of what they have learned, not just master a few techniques in the short term. The principal aim is to stimulate students to take the initiative, to actively work on ideas, not simply assent to them as they transcribe your notes. (See also *Assent – Assert*, p82.)

The tactics listed in this section apply equally well to tutorials, so you may wish to try them in a tutorial context before using them in a lecture. Following the tactics, other issues are discussed, such as structuring lectures, handling screens, how to remain fresh when repeating a lecture, and handouts.

Lecture Structure

'An intrinsic aspect of the method at all levels, elementary or advanced, is to concentrate attention on the definite, the concrete, the specific. We all have an innate ability to generalize; the teacher's function is to call attention to a concrete special case that hides (and, we hope, ultimately reveals) the germ of conceptual difficulty.'

(Halmos, 1985, p272)

Some lecturers structure their courses like serial stories. Their lectures are sections from notes which look very much like a textbook (indeed that is how most textbooks arise, from lecture notes modified over several years). The serial lecturer marks the point where the lecture ended, and takes up at the same point (or backtracks a short way) next time.

Other lecturers use a different structure. In the last few minutes of each lecture, they hand out notes and pre-topic-tasks for the next lecture. The following lecture then begins with questions from, or posed to, the students about their work on the tasks. The content of the subsequent exposition then depends on what the students ask about, and how they respond to the questions posed. Such a structure is a merging of traditional tutorial and lecturing formats, and is more suitable to small classes than very large ones. However, variants of this structure can be used, in conjunction with some of the tactics proposed in this section, for any size of group (particularly *Talking in Pairs*, p49, and *Muddiest and Most Important*, p51). See also Boelkins and Ratliff (2001).

Some lecturers structure their approach to each topic by introducing a problem whose solution spurred on the development of that topic, or could have done so. For example, probability arose partly in response to of the difficulty of fairly dividing the stakes when a gambling game was interrupted; group theory arose partly in response to the use of symmetries in solving differential equations, but could have arisen from situations such as bell-ringing (see Budden, 1972, for a variety of examples). They then either get students to work on problems associated with particularly informative examples, or they develop those examples themselves. They might work intuitively at first, giving their students a mental picture of what is going on before formulating formal definitions, relating these to the examples, and then stating and proving any relevant theorems. Where rigour and proof are particularly important, there are tactics for keeping track of the relationship between the intuitive and the formal (see *Tactics: Tracking The Intuitive and The Formal*, p53; *Catching 'it'*, p95; and *Framework: Concept Images*, p190).

Some lecturers try to vary their approach so that sometimes they begin with difficult examples and general statements, which they then publicly particularise and specialise before proving the original statement; sometimes they begin with simple examples and special cases and use these to build up to the full complexity of the general case; and sometimes they introduce formal definitions, lemmas and theorems before providing examples. In all these approaches, they are trying to display the different ways in which mathematicians encounter and make sense of new ideas.

Choosing only one approach and using it repeatedly is likely to habituate students to that particular format. This may become comfortable for them, but it may also mean that they are more dependent on their lecturer and less able to be flexible when they go to someone else's sessions. The same problem of flexibility versus fixity arises in the use of variables (few students, for example, are confident differentiating h or a with respect to x or t), and in the use of different notation (Leibniz or functional notation for integrals, Leibniz or operator notation for differential equations, and so on). (See *Tactics: Introducing Symbols*, p90; *Multiple Notations*, p13; *Introducing Definitions*, p24; and *Issue: Developing Facility*, p180).

Variation also helps to retain students' interest, so it is important that you vary not only your approach but also the pace of your lectures, and the types of interaction you encourage your students to engage in during them. Just as a picture put up on a wall disappears into the wallpaper in a few weeks, so a constant pace, form of interaction, voice tone and density of concepts can lead to students tuning out and, at best, responding mechanically.

Some lecturers use two overhead projectors (OHPs), or two boards. On one they record intuitions and conjectures, while on the other they record formal definitions and proofs. Others use one for pre-prepared notes and the other for expanding on details, working through examples, and so on. (See *Tactic: Tracking The Intuitive and The Formal*, p53).

Whatever overall structure is used, within each session there are many different tactics that can be used in order to stimulate students to take the initiative, to work constructively *on* rather than merely mechanically *through* the sequence of ideas.

Employing Screens

There are now a wide range of different screens on which mathematics can be presented. Each has its own particular strengths and weaknesses. Especially important but rather under-used are the most ancient of them all, namely the students' mental screens.

The most important feature of any screen is not what is put on that screen, but rather how it is treated. Symbols and diagrams displayed may have rich meaning for the person who constructed them, but any audience has to make sense of them from scratch. Consequently the strengths and weaknesses of different screens are related to what opportunities they afford for student attention to be focused, and for mathematical thinking to be stimulated (see *Tactic: Say What You See*, p20, 30, 91). Before considering specific screens, some general remarks about what is likely to appear on them are perhaps appropriate.

Diagrams and Symbols

There are many similarities between diagrams and symbols: both require experience in order to be able to read and interpret them easily. There are also some differences: symbols tend to be used in the same way in different contexts (though by no means always), whereas, while there are often conventions within an area (for example in graphs of functions an open circle is used to indicate a non-attained end-point) each diagram is, on the whole, unique. Whereas strings of symbols have a clear sequence, and can be expanded unambiguously in symbols or words, diagrams have implicit and invisible sequences relating cause to effect. Symbols can be read as having particular values, but usually signify a general relationship, whereas diagrams look particular but are supposed to imply a more general result as well.

Diagrams

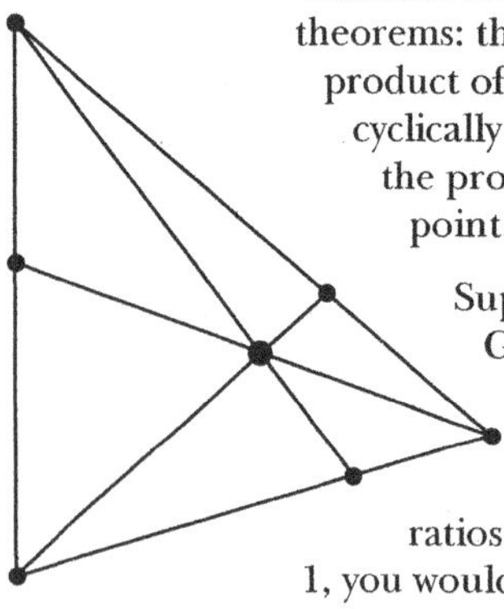

Consider this classic diagram. It could be used to illustrate two different theorems: that if three medials intersect in a common point, then the product of the ratios in which they divide the opposite sides, taken cyclically around the triangle, is 1 (Ceva), and the converse, that if the product is 1, then the medials must have a common intersection point.

Suppose you wanted to draw such a diagram in a Dynamic Geometry package, where the consequences of each theorem show up as dependencies in the construction. You would have to have two different constructions, even though the final diagrams would be identical. Thus, starting with the ratios marked out along the edges such that their cyclic product is 1, you would find that the three medials actually do intersect at a common point, whereas, starting with the three medials through a common point, you could find the three ratios and discover that their product was indeed 1.

Because of your familiarity with the result, you are aware of dependencies within the diagrams, and of the theorem and its converse, but are your students aware of this? In a diagram used to illustrate the convergence of a bounded monotone sequence, your students may be confused by the fact that it is manifestly clear from the diagram that the sequence has to converge, but blissfully unaware of the dual perception of the sequence either unfolding term by term in time, and so, by the conditions, forced to converge, or already present and complete but because of the conditions, must be certain to converge? In a diagram used to illustrate change of basis, do students have a sense of a fixed basis being used to describe alternative bases, and that both bases are equally valid and give rise to their own coordinate system? (This can be a particular problem when teaching the theory of Relativity, which depends crucially on such an understanding.) In a diagram showing the mean and variance of a distribution, are students aware of how these would change as the distribution changed, and what percentage of the distribution has to be outside one standard deviation of the mean?

Each diagram is a particular example that is intended to be generic. But in order to be perceived generically, the viewer has to be aware of what is allowed to change, and what is forced to remain invariant amid that change. They have to know what needs to be stressed, and what is background. For example, with Ceva's Theorem, either the common intersection point can change (anywhere on the plane), leaving the cyclic product of ratios will invariant, *or* the ratios can change subject to the cyclic product remaining invariant, forcing the medials to have a common intersection.

A similar phenomenon can be found with almost any other diagram if that diagram is to serve a useful function. To be perceived as generic, some features in a diagram are permitted to change, and some have to remain invariant. Furthermore, as with Euclidean constructions, there is often a preferred order for drawing the diagram, either to show increasing detail, to mimic some unfolding process, or to display logical dependencies. By being aware of these when you are drawing or using a diagram, you can make the diagram much more effective for students. If the diagram is pre-drawn, then extra care may need to be taken so that students appreciate what is generic about the diagram and what is particular. If you are drawing the diagram, then try drawing it slowly and discussing dependencies as you go, rather than quickly drawing it and then discussing the static result. Bear in mind that students may be trying to copy the diagram at the same time as trying appreciate what it is indicating. If they end up simply copying it, rather than reconstructing the structure, the diagram may not be of much use to them (see also *Tactic: Drawing Diagrams*, p92).

Symbols

Mathematics is most efficiently communicated via symbols, but written mathematics needs lots of words around the symbols in order to help the reader make sense of it all. In lectures, you can use these words to stress certain aspects and to indicate what is generic and what is particular. While some students can manipulate strings of symbols according to prescribed rules without needing more than a superficial understanding of what the symbols actually mean, others have to understand what they are saying in order then to consider whether it makes sense. In any case, whether or not your students would be able to complete their assignments without it, they will not make good mathematicians unless they have a deeper understanding.

One of the purposes of putting strings of symbols in front of students is to familiarise them with reading and interpreting them. Another purpose is to encourage them to use symbols to express what they are thinking. So sometimes in a lecture you can treat a sequence of displayed symbols as an exercise in decoding, making sure that they have their own copy of the symbols before you embark; other times, the symbols can emerge as an expression of what you are thinking in public, so that students see you using symbols to express your thoughts. Of course, only the most accomplished experts can write mathematics perfectly the first time. If students can sometimes see an expert editing and revising so as to more precisely capture what they are trying to express, it can give them an insight into what they may have to go through. Many students seem to think that mathematical symbols should emerge perfectly formed from their pen as it writes on the page, and so they are depressed and frustrated, even demoralised when they find expressing themselves in symbols so hard. (See also *Tactic: Introducing Symbols*, p90)

Using Screens

There are many kinds of screens, mental, electronic, and physical. Much of what applies to one applies to them all. In particular, you can usually get less on a screen effectively than you imagine, though a blackboard can accommodate the most.

Blackboards

A suite of blackboards enable an entire proof (well, usually) to be put on the board, and then discussed as a whole. But it is time consuming to write out complete words and sentences, so it is often easier to write in succinct symbols and then talk around them. If the students see their task as merely to make a copy, then they could perhaps do this in their own time later (see *Handouts*, p64). However, if the purpose of displaying a proof, a worked example, or particular cases of a theorem is in order to get a sense of the whole, to appreciate some fine points or clever ways of thinking in the argument, or to be exposed to the structure of mathematical proofs, then more has to be done than merely presenting the proof on the board.

Firstly, in order to discuss a proof as a whole, students have to finish doing whatever transcribing or reading is necessary. Then they have to have their attention directed to the parts that you consider important or noteworthy. Do not assume that just because something strikes you, it will strike your students similarly. Remember that they do not have your wealth of experience. Words, voice tones, gestures, and posture, are all involved in indicating what you are stressing, what you are attending to, what you think is important. Relationships and connections can be indicated physically in a way that is very difficult in speech, since it is steadfastly sequential in time. However, students need time to capture that in their notes!

Overhead Projectors

The most frequent misuse of an overhead projector is to stand between the audience and the screen. The next most frequent is to obscure the projector with a shoulder while writing or pointing at the acetate. It is essential to check sight-lines before the first session begins in a new room. The best arrangement is to have the screen(s) in the corners of the room angled slightly, but this is not always possible.

OHPs show very little at one time. If acetates are pre-prepared, the temptation is to put too much on each screen, and then to work through them faster than anyone can take notes. In any case, the point of displaying things is to demonstrate how to think about what is displayed, how to work on ideas. If written 'live', then details are likely to be omitted through lack of space, while shorthand notation may be confusing for students trying to make sense of their notes later. Rolling the acetate forward and back is just as hard on students as interchanging slides quickly. If you are going to write on the OHP, make sure that it is positioned so that students can see, or make sure you frequently step away so everyone can see the screen. Stepping away introduces a slight pause, during which you can consider what to do next while your students catch up and think about what you have said and done.

Whenever you want students to focus on some detail, make sure that you are pointing to it explicitly, either physically with a pointer, or verbally. You want the students to pay attention to what you think is important, so do not move a pointer rapidly over the display or jab at the slide with your finger. Rather, place a pointer on the acetate; slide it slowly to its next position, and leave it there. Do not point with your finger: either your hand or your shoulder are bound to obscure the screen for some students. Make sure that everyone, including those who look down to take notes of what you are saying, can see where you are pointing.

Though much less can be presented on an OHP than on a board of the same size, the effect is like that of binoculars. It can be used to focus attention, to frame and hence cut out extraneous features. Thus an OHP is excellent for summarising, for presenting an outline of what you are going to do, and then later of what you have done. It is not terribly useful for providing full details of arguments or of worked examples. Giving students copies of the slides with appropriate spaces for them to add their own notes works well for some lecturers (see *Handouts*, p64), but do not forget that students need time to absorb a flurry of new ideas, or even old ideas in a new order or new notation.

It is much more effective to prepare headings, and then to use another OHP or a board on which to work through the details (see *Multiple Screens*, p47). Of course, you are likely to make the odd mistake, but that heartens students, and watching you respond to a mistake or difficulty also gives them some insight into how mathematicians actually think. By all means have notes with you so you can get out of difficulty, but remember that the purpose of a lecture is to engage students' thinking and attention, not to show how much you can 'cover'. (See also *Tactics: Using Multiple Representations*, p19; *Intensive and Extensive Definitions*, p25; *Getting Students Used to the Idea of Proof*, p53.)

Epidiascopes

These are excellent devices for taking text (including diagrams) on paper and displaying it on a screen, saving the need to copy or enlarge it for display. Many institutions have one or two, but they are not often used. They remove the necessity for making slides, but they suffer in the same way as OHPs in that there is a strong temptation to try to present too much on a screen at once.

Fixed Video Cameras

Increasingly popular as replacements for epidiascopes, but with much greater potential, are fixed video cameras on a stand pointing down at a page. Not only can these display text and diagrams on a screen, but they can be used to display three-dimensional models and other objects. The object can be turned while you talk and point, to give students a sense of the object in all three dimensions. Furthermore, the focus can be altered so as to help students appreciate depth. Apart from their greater flexibility, these 'screens' are much like OHPs and computer screens: an overly full screen is very hard to make sense of from the audience's point of view. In addition, relatively slow movement of the real object may appear exaggeratedly quick on the projected screen.

Computer Screens

Multimedia offers tremendous opportunities for presenting mathematical ideas. Software for lecture or tutorial presentation, or for students to use themselves, and access to the web means that students can be shown real-time results of complex calculations, geometrical constructions, and mathematical models. They can watch an object being translated, rotated, reflected, and scaled; they can see iterative approximations built up and recursive schemes worked through in real time; they can see the results of better and better approximations being found; they can alter parameters and observe the effects; they can even try to 'teach' a computer to do calculations, and find out what can go wrong!

There are some caveats, however. It is vital that students appreciate what it is that the computer is doing. While multimedia offers tremendous opportunities for displaying phenomena which can best be interpreted using mathematical vocabulary, and which can be seen as illustrating important and difficult mathematical ideas, students need to have their attention drawn to what is invariant and what is permitted to change, either by the lecturer as changes take place, or as an explicit task (see *Tactic: Invariance Amid Change*, p32). Software can be used effectively for lecture or tutorial presentation, as long as students are supported and stimulated into working on what they see; software can also be for students to use themselves, perhaps through prepared worksheets that the students can modify. If students are given recipes to follow, then the session is likely to be as successful as old-style physics and chemistry 'experiments': students go through the sequence, but emerge little the wiser. Similarly, presented with a multitude of hyperlinked web pages, students may feel forced to keep clicking without stopping to make sense of individual frames. If students are provided with tools and a starting point and set the task of constructing an object having certain properties, then they are more likely to enrich their mental images and their understanding of the topic.

The notion that the use of multimedia simplifies concepts, diagrams, or arguments for students is largely a myth. A complex medium often gives rise to a complex result, unless a lot of effort is invested in making the presentation clear. This means even more work for students to make sense of it all (unless you are simply after a 'wow' effect). Furthermore, the more complex material is in its form and style of presentation, the more frustration you are likely to experience in trying to adapt someone else's materials to your own use. What for an expert is enlightening because it provides a focus for unifying disparate ideas can be a complete mystery for students.

Smart-boards

There is growing use of smart-boards, which act both as a touch-sensitive projection of a computer screen, and as a board on which to add remarks. The computer responds to touch on the screen by a finger or pointer much as it would to a mouse. Students who see the presenter interact with the screen get a better sense of involvement than when a sequence of clicks is made by an almost invisible mouse.

Mental Screens

The most important screens in a lecture theatre are the mental screens of the students. If everything happens so quickly on a physical screen that they cannot transfer it to their own mental screens, then significant potential is being thrown away. If, every so often, the lecturer pauses and invites students to draw their own diagram, or to imagine some things happening mentally before drawing a diagram for them, then students are more likely to make sense of subsequent diagrams (see *Tactics: Say What You See*, p20, 30, 91; *Invoking Mental Imagery*, p55). If students are invited to predict what the outcome of a transformation or calculation will be before being shown the result, then they are far more likely to recognise that their initial conjecture needs to be modified and so to recognise what is being offered by the theory than if the examples and counter-examples flow past in an endless stream. If students are invited to listen to and contemplate the outline steps of an argument (without taking notes) before being presented with the details, they are more likely to make sense of the whole than if they just encounter the details (see also *Issue: View – Serialist or Holist*, p165).

Multiple Screens

Using more than one screen can be a real advantage, both to students and to the lecturer. For example, one screen can be used to record intuitive ideas and intensive definitions, while the other can be used to record extensive definitions and formal theorems (see *Tactics: Tracking The Intuitive and The Formal, p53*; *Intensive and Extensive Definitions*, p25). Alternatively, one screen can be used for examples of mathematical objects or for worked examples and the other can be used for generalisations and abstractions. The advantage to students is that they are helped to distinguish between the particular and the general, and between the intuitive and the formal. The advantage to the lecturer is that they too are encouraged to pay attention to the processes of formalisation, generalisation, and abstraction. There is less danger of being caught up in the flow of pre-digested and sorted-out ideas, and more hope of being reminded to expose students to the transformation from intuitive to formal, with all its surprises and requirements.

Tactics

The tactics described here often suggest getting students to interact with you, perhaps to ask questions or to volunteer observations. You may feel that this is too difficult in a lecture room of 30, 100, or even 300 students. However, each of these tactics has been employed fruitfully with audiences even larger. You cannot always get everyone to say something, but you can set up conditions in which individuals, pairs, and triples can do useful work. You can also get some feedback about what students are thinking, and, most importantly, you can display techniques for making sense of mathematics that students can use among themselves without needing you to be there.

When you try a tactic for the first time, do not be put off if students do not respond fully at first. They too have to get used to new practices. If nothing happens at first, do not blame yourself or your students; work at changing the implicit contract between you and the students instead (see *Issue: Implications of Student and Tutor Expectations*, p166; and *Tactic: Being Explicit About The Enterprise*, p56). Do not expect to know exactly what they are all doing, but put your trust in the students wanting to take every opportunity they are given to make sense of the material. A lot depends on your general demeanour: whether you come across as genuinely concerned about their learning, or as more concerned about covering the material.

Tactic: Treating 'Screens' as Phenomena

Tactic: Instead of thinking of acetates, computer output, or text as descriptions of mathematical thinking, try thinking of them as objects, as phenomena on which to work. Having displayed a slide or a page, treat it as an object to be analysed, even criticised. Use some variant of the tactic *Say What You See*, (p20, 30, 91) to help students become aware of details. Get students to try to (re)construct a story which makes sense of or enables them to redraw a diagram or symbol sequence for themselves. Hide what was shown and get students to reconstruct it for themselves. You may find that how they pay attention to what you show on a screen changes as you use this with them over time!

As you work through text and diagrams, draw attention to the part you are looking at, using a pointer lying still on the projector. Do not just read through the text as if it were self-explanatory. Rather, as you work through, comment on features such as what makes a particular construction clever, or the need for detailed calculations (which can be returned to later), the structure of the argument, or how a particular theorem or technique is employed.

Tactic: Advance Organisers and Summarising Organisers

Teachers in universities and colleges are increasingly being called upon to state the learning outcomes of each session. It is based on a useful notion, that students like to have some sort of an overview of where they are going and what will be happening, and deserve to be told. Some students find it difficult to concentrate when they find themselves having to follow the whims of the lecturer's thoughts without having their perspective. Thinking about what a session is expected to achieve is one thing; writing it all down so that someone else (an inspector, colleague, or student) can make sense of it is quite another (see also *Learning Outcomes*, p127).

Tactic: Provide a succinct statement of what you expect the session to be mainly about (for example a specific proof, a particular idea, or a clever technique). If you can arrange to remember before the session finishes, remind them at the end of the session and perhaps comment on something pertinent or particularly important.

For example, before rehearsing and developing integration or differentiation, ask students to note down, or perhaps even say to a neighbour, what they think the term means. You can then say that the session or the course is about understanding when these are possible, in other words finding meanings for 'integrable' or 'differentiable'. Some lecturers like to invite students to think (briefly) about the notions that will be developed formally, so that they are oriented towards and appreciative of the problems that might arise (see for example Legrand, 1993).

Tactic: After a period of time, try getting students to write down on a card or slip of paper what they thought was the most important aspect of the session, or what they thought it was about, and to leave these behind at the end. The cards may make interesting reading.

The principal aim is to provide students with time to think back over the session and see it as a whole, because the moment they start to pack their things away their minds will have moved on to the future, and their thoughts about the session will have evaporated. This tactic helps students as well as informing you.

Tactic: Talking in Pairs

Tactic: When you catch yourself saying a complicated sentence, or describing a complex idea, stop and ask students to say to each other what you have just said. When you describe something using an image in your head, make sure that you use the imperative (see *Tactic: Say What You See*, p20, 30, 91), then stop and ask students to describe to each other what image this conjures up for them.

It takes only a matter of seconds to stop and ask them to interact, though at first they may be reluctant. Do not move about or look at your notes, but rather look expectantly at them or, better, carry out the task mentally yourself.

Once they start talking, the noise level will rise and you will not be able to regain their attention until it starts to subside again. When it does, you can ask if they have any questions, or you can simply carry on. Often just the opportunity to say something to someone else can be beneficial, as well as changing the energy in the room. In a tutorial, as you and your students gain confidence, you can ask if anyone struggled, or whether they would like to try to say something publicly, and then you can use this to develop discussion (see *Issue: Discussing Mathematics*, p176).

In 1998 the Strathclyde Engineering Department began experiments in which students responded to lecturer questions by entering their choice of answer on a personal handset like a calculator, which then automatically displayed summary statistics. This gave the lecturer the opportunity to get students to try to decide which answer was correct, and why. They could discuss it at first locally (for example in pairs) and then express a view to the whole class. The feedback informed the lecturer as to whether to go back over an idea or to press ahead. They reported that dropout and failure rates declined significantly.

Tactic: Pausing

In a lecture, the presence of students provides an impetus for you to make fresh contact with the world of mathematics, and to attract students to enter that world with you. But moving about in a strange world takes time: a good guide pauses to let people savour something or catch their breath. If you notice part of the way through that there is a general round of coughing and shifting of posture, then it may be that you are not managing to hold your students' attention, which is no wonder as few people can sustain attention in what someone else is saying for more than about 20 minutes. It is vital, therefore, to provide time for your students to think and assimilate, especially near the end of a session, because the moment the session ends everyone's attention will immediately be directed elsewhere, and by the time students re-work their notes (if ever), the freshness of the session will probably have evaporated, leaving them with notes that look like a poorly written textbook.

Tactic: Just after saying a complicated sentence, or at the end of a proof or an example, pause and look at the board or screen, displaying by your posture and gesture that you are reviewing what has been done.

You cannot guarantee that the pause will be used effectively, but you can remind students every so often of what they could be doing during the pause. For example, you might suggest that they 'look over' the example (or the theorem) and try to tell themselves what it is about (see *Tactic: Talking in Pairs*, p49), or that they try to form an image or even write down a particular or peculiar example for themselves (see *Tactic: Particular – Peculiar – General*, p88). Then, instead of starting straight in with the next item, you can ask whether the pause has generated any pertinent questions or revealed any stumbling blocks.

If during your pause you pace about or fiddle with your notes, you indicate by your behaviour that you are simply waiting. If, instead, you stand still and do yourself what you have asked your students to do, you will be more likely to stimulate them into using the pause effectively.

Tactic: Emphasising One Thing Per Session

Tactic: Choose one thing per session, and, when you get to it, pause, say something like, 'if you remember nothing else from this session, then try to remember this ...', pause again until you have their attention, then state it succinctly, preferably more than once.

Be careful, however; you do not always want to have students remember what you say: off-the-cuff remarks may have unintended effects. A colleague once referred to 'some old Greek' in connection with a theorem because the name slipped his mind at the time. In the exam it came back to him as 'some old Greek's theorem'!

Tactic: Making Memorable Remarks

Tactic: Labelling important theorems and techniques memorably can be a great help to students.

For example, humourously referring to the Orbit-Stabiliser theorem in groups as being due to professors Orbit and Stabiliser can make the theorem stick in students' minds. (Then again, it might also be misinterpreted by someone taking it too literally, as with 'some old Greek's theorem'!)

If you can capture the essence of something in a snappy slogan, then by all means try it out on students. Beware, however, of the temptation to construct mnemonics, for sometimes the mnemonic becomes an obstacle. For example, the CAST rule for the signs of trigonometric functions can become so deeply embedded in students' minds that they cannot get beyond it to a sense of where the different functions are positive; they become stuck with having to use it each time to work out the signs. Similarly, some students want a mnemonic in order to remember SOHCAHTOA, itself a mnemonic for the trigonometric ratios.

Tactic: Muddiest and Most Important

Tactic: Every so often, ask students to write down what they find the muddiest or the least clear concept, example, definition, theorem, or technique. You can have these handed in, ask for someone to offer one to the whole group, or suggest these are taken up in a tutorial. After a while you may begin to find that they are more forthcoming in asking questions without this prompt. (See Angelo and Cross, 1993, p160–163 for variants.)

Tactic: Every so often, at the end of a session, ask students to write down on a card what they think were the salient points in the session. At the same time, you can also write down what you think, together with what you predict they will write.

You may be shocked at their responses, at how little they seem to understand or appreciate. If you assume that students instinctively understand what you have told them, that they naturally pay attention to the same things as you do, then you are likely to be surprised. Achieving such a state of affairs takes time and experience. The point of this tactic is to gain insight into what students are stressing and thinking. Do not assume that, just because you have said something to them, it is either known or understood. Students may not be in a sufficient state of preparedness or competence even to be able to hear what you say, much less integrate and appreciate it.

Tactic: Tracking The Intuitive and The Formal

George Polya (1957, p72) suggested that to understand mathematics is 'to prove formally what is seen intuitively, and to see intuitively what is proved formally'. Students bring mathematical knowledge to college, but sometimes it is repeated as if they had never seen it before, whereas sometimes it is assumed without explicit reference. Sometimes it is used as the basis for further development (as in linear algebra), and sometimes it is used as the intuitive base for reconstruction from a rigorous foundation (as in early analysis). Students often complain that they do not know what they are allowed to assume and what they have to prove. Using two different screens can be very effective for this.

Tactic: Use two screens and/or two sides of the room, one for discussing intuitive ideas, and the other for formalising those ideas. Stand in one place to express formally something that students have some sense of intuitively, then change sides and screens to discuss some of the implications.

The discussion in Chapter 1 around the tactic of *using 'strange' examples* (p25) is strongly supported by using this approach to help students keep track of the difference between the intuitive and the formal.

A similar tactic can be used to emphasise the way in which particular cases can aid appreciation of the general case, and in which the general subsumes and generalises the particular. One side of the room and/or one screen can be used for particular examples, while the other side and/or screen can be used for generalisations. This applies also to proofs. By separating particular and general, the two can be developed in parallel, step by step, either with the particular first and then the general, or the general and then the particular. Of course, there will be steps which seem tedious and do not require following through in detail (rearranging algebraic expressions for example), but for some students these could be valuable to do in their own time. The delay in having to move from place to place is actually valuable for preparing your students for the kinds of remarks you are going to make next.

Tactic: Presenting the Essence of a Theorem

What are theorems really for? What role do they play? What role do proofs play? Many students do not really appreciate what theorems tell them or what they are supposed to do with a proof. Are they supposed to understand them, memorise them, reconstruct them, learn from them?

For example, some theorems are statements of what remains invariant when something else is permitted to change (applying a linear transformation to a set of data applies that same transformation to the arithmetic mean; adding a constant leaves the derivative of a function invariant). The theorem indicates the scope of allowable change but, for various reasons, students may not appreciate the scope of its generality. One strong reason is the use of *a, an* and *any,* which are often ambiguous or at least confusing as to when they signal the particular and when the general (see also *Tactic: Expressing Generality,* p14).

Tactic: Try emphasising the scope of a generality within which something else is invariant, so that you can stress this feature for students. Constructing examples, and getting students to construct examples that illustrate this scope, can also be useful.

For example, 'imagine a differentiable function specified on an interval $[a, b]$, and look at the slope of the chord between the end values on that interval. Somewhere in the interior of the interval there lies a point at which the slope of the curve is the same as the slope of the chord. Now allow your function to become very wiggly, and again even wigglier, as long as it remains differentiable. It must still be true that, somewhere in the interior of the interval, there must be a point ...'. (See *Difficulties with Logic*, p27; *Tactics: Student Generated Examples*, p16; *Distinguishing Common and Technical Meanings*, p27; and *Theme: Invariance Amid Change*, p192.)

Other theorems show that, in a specified context, one property implies another. Some stronger theorems show that several different-looking statements or conditions are actually equivalent (often proved by the apparently circular proof structure of P implies Q implies ... implies P). Finally, there are theorems which establish that a particular technique is guaranteed to work under stated conditions.

Tactic: Try going back over a proof, pointing out the structure or role of significant ideas or moves. Include with this some commentary on the role of the theorem in the topic. Later, ask students to decide what role the theorem plays in the overall theory. Ask them this every so often, hopefully weaning them off their dependency on you to ask the question.

As with any of these tactics, overuse can rapidly turn into abuse, but not to use it is to squander potential for students to appreciate mathematical thinking as well as to learn mathematics more efficiently. (See also *Tactic: Directed – Prompted – Spontaneous*, p91.)

Tactic: Getting Students Used to the Idea of Proof

Some proofs are informative because they explain why something must be (or cannot be) the case. Other proofs are highly elegant, but not terribly informative. Some expose connections between otherwise apparently disparate topics, while others are technical results which are useful in order to short-cut work later. Do students appreciate these different roles? If proofs are for convincing people, why do mathematicians appreciate several different proofs of the same result?

Proving does not come easily to most students, and simply seeing lots of proofs go by may not be sufficient to enable them to appreciate the role of proofs, or to construct them for themselves. Although mathematicians see proof simply as a mathematical form of rational reasoning, sequential logical reasoning is not easy to sustain. Students need to appreciate what a proof is intended to achieve, and what the rules are (see *Tactic: Scrambled Proof*, p94).

To develop a mathematical proof it can be helpful to think in terms of first convincing yourself, then convincing a friend, and only then convincing a reasonable sceptic (Mason *et al.*, 1982). This tactic was used by Galileo, who wrote his proofs as plays involving himself, an advocate of the opposing theory, and a reasonable sceptic. By cultivating a conjecturing atmosphere in tutorials, you can encourage your students to work in a similar way when you are not present. In this way, you can help your students to behave like expert mathematicians.

Tactic: Try constructing and explicitly using a proof template for a sequence of proofs in your lectures. This will mean that you can, at various times, pause and draw your students' attention to the important elements: assumptions, steps using agreed results, and checking that the conclusion matches the stated claim. Some lecturers like to use a structure such as, 'What do I Know? What do I Want? How can I get from one to the other?' as a framework of questions to ask themselves publicly as they are presenting proofs. This provides a format for students to follow when they get stuck (Mason *et al.*, 1982).

Uri Leron (1983, 1985) recommends a top-down approach: 'If I knew this and this, I would be done; now look at each requirement in turn.'. Other people map out the main components of a long argument, even using a tree-diagram with nodes representing sub-arguments. Then they pause every so often and relate what they have just gone through to the map. On completing the details of the proof, it can be helpful to then review the overall map again. Using two screens can help with this.

Tactic: Ask yourself how you know what to do at each stage, and then try being explicit about your choices. Did you work top-down, or bottom-up, or some mixture? Highlighting what you think about at each step is perhaps the most valuable support you can give to your students. (See *Tactic: Making an Example be Exemplary*, p29; and *Issue: What is Exemplary about an Example?*, p173.)

Points to be considered may be common to several proofs, such as:

- in analysis, showing how the clever choice of δ makes the final inequality exactly what the definition requires (or, conversely, showing how you convert the first version of the argument – where you end up with extra multiplicative constants for ε – into a formal proof which ends neatly with the required ε on its own);
- in linear and abstract algebra, interpreting technical terms such as *linearly independent* or *subgroup* as symbols and constraints which can then be manipulated.

Proofs often depend on seeing a connection, and seem entirely obvious and natural once you have seen it. However, if you have not seen the connection, progress is likely to be blocked. For example, the following remark appeared in the first general proof of a result (another technical proof had been previously published).

'The expression ... is clearly a discrete convolution, which immediately suggests an approach using generating functions. It is convenient to change notation beforehand ...'

(Ridley, 2000, p358)

What was it that triggered the recognition of the discrete convolution, probably intertwined with the recognition that generating functions might be useful? How can students be helped to make such connections for themselves? At the very least, when presenting the proof to students, one would want to point out what it was that triggered the recognition of the discrete convolution, how this led to a useful change of notation, and why convolution and generating functions are closely linked.

Tactic: Commenting on Proofs

Tactic: Having presented the reasoning and algebra of a proof, pause and go back over it (once students have copied it down) pointing out at each step what triggered the next step, how the proof is similar to or different from other proofs, what was really insightful and what was routine calculation. In a tutorial context, you could ask students to select what they thought was the key moment of insight, and to discuss it briefly in pairs. You could then get them to contribute to plenary discussion or simply present your own perspective. (See *Tactic: Talking in Pairs*, p49.)

See *Tactic: Being Explicit About the Enterprise*, p56

Tactic: Invoking Mental Imagery

Just over half of the population responds to the language of *seeing* as a description of how they understand ('I see what you are saying'), while around a third respond better to *hearing* as their metaphor ('I hear what you are saying'). Those who like to 'see what is going on' respond well to having specific objects to manipulate which are confidence inspiring for them, whether physical objects, diagrams, or symbols forming 'concrete examples' with which they are familiar. Those who prefer hearing may respond better to verbal descriptions, commentaries, and instructions to employ their mental screens (for example 'imagine a ...'). (See *Tactic: Invoking Mental Imagery*, p55; and *Framework: Manipulating – Getting-a-Sense-of – Articulating*, p187.)

Mathematical imagery does not arise spontaneously. Indeed, many students seem not to be aware that they have any control over the images, connections, diagrams and techniques which 'come to them' (or fail to come to them) in association with a technical term. Furthermore, our image-rich media culture reinforces the mistaken notion that images have to come from other people as complex pictures. One way to counteract these trends and to support mathematical thinking is to invoke mental imagery explicitly. There is an endemic tension, here because, although no two mathematicians 'have the same imagery', there is much more in common between images of standard concepts than students often realise.

Tactic: Before announcing a theorem or drawing a diagram, use imperative language ('imagine a cubic curve', or 'imagine a chord drawn between two of the points') to encourage students to construct their own mental image. If they seem to be faltering or struggling to keep control of all the components, getting them to describe to each other what they are seeing (without using their hands or drawings) can help them to gain flexibility in moving from words to diagrams and back (see *Tactics: Talking in Pairs*, p49, *Say What You See*, p20, 30, 91).

Tactic: Before manipulating a line of symbols, pause and invite your students to imagine the manipulation being carried out, as a means of stimulating them to strengthen their ability to mentally manipulate symbols.

The idea is to get them to work actively *on* what you present, not simply allow it to wash over them. This means you have to slow down but, as with all these tactics, they will become much more efficient in the long run and you will be able to progress more quickly.

Tactic: Being Explicit About the Enterprise

'To teach "real variables" ... begin with interesting facts instead of obvious axioms and artificial definitions. Begin with examples of strange numbers, strange closed sets, and strange functions. Look at examples and counter-examples, look at continuous and discontinuous functions, look at uniform convergence and its opposite. Make such things friendly and familiar ... and then *raise the question of what could have been a good starting place to get here from.'*

(Halmos, 1985, p272)

In some courses (for example analysis) students see intuitive ideas being formalised. It is possible to be explicit about this process, for example by standing in one place in the room for talking about intuitive notions, and in another for talking about formalising them (See also *Lecture Structure*, p40). For example, some lecturers stand on one side of the room to select 'the epsilon', and the other to search for and locate 'the delta', before returning to the epsilon side to construct their final argument. In other courses, students meet ideas which generalise familiar ideas (linear independence, dimensions of a vector space, vectors) and they are expected to work with formal definitions. The most difficult type of course is one where students meet new ideas in unfamiliar contexts (topology, algebraic topology, group theory). Some students may not recognise the difference between the different types though!

Tactic: Articulate to yourself what the principal mathematical enterprise is in your course: perhaps encountering new ideas; formalising old ideas; exploring consequences of axiom systems; developing techniques for use in other courses; turning the geometrical into algebra; or turning the algebraic into the geometric. Then mark your notes with places where you can refer to that enterprise and be explicit about it, in order to help students distinguish the wood from the trees, the overall from the detail.

Simply presenting carefully constructed examples from which you generalise, or presenting formal definitions, lemmas, and then theorems and their proofs, do not constitute being explicit about the enterprise. Being explicit means catching your own awareness that what you are currently doing is an example of the how mathematicians work, and being explicit to students about it. (See *Tactic: Intensive and Extensive Definitions*, p25.)

Tactic: Exposing Inner Monologues

Tactic: While you are carrying out standard techniques in public (on a board or at an OHP), verbalise what you are thinking as you do it: 'I usually start with the inside bit ... then I take this over here ... now what do I want? ... Well, I know that ... so ... ', and so on.

At the very least this will awaken some students to the fact that you have an inner dialogue with yourself while performing a technique. This gives them some idea of what your own mental screen looks like as you are working. There is no need for them to take notes, so it is, for once, an advantage that you will probably go fairly quickly through it. By doing this, your students get to hear that their tutors have inner chatter, and to hear the shape of those monologues. They can also get a sense of what it is like to be skilled in using the technique. At the Open University we call these sessions 'technique bashing'. They last only 20 to 30 minutes, but are found to be very valuable by students who have been working on a topic. It provides a change of energy, and an opportunity to see an expert make mistakes, check their work, and expose their inner thoughts.

Tactic: Being Human

'To see an expert make a mistake or admit ignorance, and then battle his way to the truth, can be an eye-opener.'

(Halmos, 1985, p135)

If students only ever see perfectly articulated mathematical expressions flowing from the chalk or OHP pen of an expert, they miss out on an important dimension of mathematics: that it takes many drafts and many attempts to reach a clear articulation of an idea or proof. Without suggesting that it is worth setting yourself up for a fall, it is nevertheless good for students to see you struggle sometimes. It is more likely to happen in a tutorial, where you have to respond to student questions, but it can be useful to get a colleague to set you a question to do sight unseen in front of students. Of course, if you get into very hot water, and cannot see your way clear, you can suggest that students work in pairs while you work on it, or you can leave it (perhaps with a conjecture) and return to it another time.

It takes courage to open yourself to making mistakes, but if students see that you too can err, then they are more likely to appreciate the way that you lay things out clearly for them.

Tactic: Take (or make) opportunities to work 'live' in front of your students occasionally so that they can see you struggle, follow dead ends, and massage and modify conjectures and articulations until you reach a satisfactory result. They may not follow every detail of your thinking, but at least it will show them that mathematics is done, indeed discovered and constructed, by human beings.

Tactic: Learning from Experience

One thing we do not often learn from experience is that we usually do not learn from experience alone. Thus, when we come out of a session wishing we had done something differently or recognising that we could have done something better, it is easy to think, 'I won't do that again', and yet, a year later, find ourselves doing something very similar! For example, after a session, we may wish we had spent more time on a diagram or example rather than getting caught up in the details of some argument, or thought of a better example to illustrate some point.

Tactic: Try making a note about what you wish you had thought to do, and adding it to a file under that topic (if it is a topic-related matter), or to a file on teaching tactics. These can then be consulted when preparing for future lectures.

It is obvious advice, but it is 'more honoured in the breach than in the observance'. Keeping a file for each topic, perhaps using the framework developed in Chapter 7 under the heading of *Concept Images* (p190), makes preparation in the future much easier and more effective than simply reading through old lecture notes. (See also Appendix B for an example.)

Other Lecturing Issues

This section addresses the problem of remaining fresh when having to repeat a lecture, deciding what sorts of handouts to provide, and encouraging exploration around topics.

When you first start teaching, your concern tends to focus on the mechanics of teaching (using boards and OHPs, setting assignments, responding to student questions, stopping on time, or covering required topics). Many lecturers are also led to learn the material in a new and deeper way, so as to be confident about responding to possible questions, and gaining facility with all the theorems and definitions. It is one thing to know results and to use them in your own research, or to have passed exams on them in the past; it is quite another to face a roomful of students. Some people write out detailed lecture notes, and then read them out and/or write them on a board in front of their students. Others practice writing out the main points and rehearse the theorems in their heads while waiting for a bus or washing dishes, so that they can reconstruct the mathematics from memory; these are but two of many approaches.

Preparing the subject matter is only part of what a lecturer can do for their students, however. A range of proposals about what else can be done can be found in later sections (see, for example, *Tactic: Catching 'it'*, p95; and *Framework: Concept Images*, p190). The immediate concern here is how to remain fresh once you are confident about the subject matter and about the mechanics of teaching. If you are bored or disdainful about the subject matter, students will pick this up, no matter how you try to disguise it. Having picked this up, they are then unlikely to respond positively to your course.

Punctuation

A lecture has many similarities with text. For example: each chapter of a book has a sense of beginning and end and so could a lecture; each paragraph elaborates on a central or core idea, and a lecture could be seen as made up of a sequence of elaborations; each sentence has a structure and that structure varies in order to maintain interest. Text is broken up by paragraphs in order to provide structure and suitable 'pausing points' to reconsider or reconstruct what has been said. Lectures can do the same. It may be necessary to inform your students of the purpose of pauses as opportunities for them to re-think what has just been said, and to remind them of it every so often, but even a short pause can be useful. It allows people to catch up, and permitting them to rest for a moment or two can enable students to concentrate for more of the lecture-hour than might otherwise be the case.

Although people often think that they speak in sentences, it is very rare in conversation (unless you are actually reading from text). In conversation we start sentences but do not finish them, switch between sentences, and change subject and object rapidly. Although a fluent lecturing style may be more like text than is conversation, and although written mathematics comes in complete 'sentences' with subject, object and predicate, the comments you make over and above what is written are likely to be more like conversation. The freshness that comes from thinking on your feet may also, unfortunately, demand that students listen more attentively because of the switches between thoughts and incomplete sentences.

Tactic: Taping Yourself

Tactic: Try making an audiotape of one or two lectures, and then listen to it while you are driving or doing something else, to get a flavour of how you sound to students, without going into too much detail.

You are likely to find that what at the time seemed like a lengthy pause was in fact very short indeed. What seemed at the time to be punctuated with a variety of types of commentary and remarks came across as nearly seamless flow! Punctuation plays an important role in exposition, whether spoken or in text. Full stops signal completions of a thought and the taking of a fresh breath, while a change of paragraph signals a new idea. Commas provide parallel syntactic structures. Each of these, as well as colons and semi-colons, have analogues in lecturing, mostly involving a combination of voice-tones, silence, gestures, and change of position. By being aware of the substructures in what you are presenting, and marking those substructures with an appropriate pause, change of tone, or explicit signal, you can make the structural aspects of your lecture much clearer, directing attention to what is most important: the mathematics.

Everyone has attended lectures in which the lecturer, perhaps through nervousness or though desire to pack a great deal into the time, moves along rapidly with barely a pause for breath. It is very difficult to keep up with a 'breathless' delivery for any length of time unless it is summarising what you already know, in which case it may not be needed anyway. On the other hand, if a lecturer goes at a very slow and measured pace, pausing to think about every single point, minds will soon wander. Establishing a 'walking' pace is important; varying that pace, and punctuating it, is even more effective (see *Issue: Time – Coverage and Pace,* p164).

Tactic: Punctuating Your Material

Tactic: Try paying particular attention to punctuating your lecture. Some important examples are the following.

Use different places in the room for presenting different kinds of ideas (see *Tactic: Tracking The Intuitive and The Formal,* p53), and use movement between positions to be quiet, or remind your audience of the significance of the new position.

Use a gesture as a device – rather than just a 'tic' – by pausing briefly as you hold the posture; mention asides that pop into your mind, but indicate in some way that they *are* asides.

Indicate in some way the beginning and end of an example, perhaps by getting your students to think about exactly what is being exemplified (see *Tactic: Advising Students How to Make the Most of a Worked Example,* p97; and *Issue: What is Exemplary About an Example?,* p173). Alternatively, ask them how they would modify, extend, or vary the example while retaining its essential features (see *Tactic: Boundary Examples,* p14, 136).

Being Interested and Stimulating

How can you be fresh and interesting when the topic is routine and elementary? If you are working on your own research problems, time spent teaching students relatively elementary mathematics can seem at best a diversion and at worst an impediment. However, it is my experience that, no matter how deeply involved in my own research I may be, working with students can be an advantage and a help, not a hindrance.

You will know from your own experience that, when you are attending a seminar or lecture course by someone who has just developed the material, there is an air of excitement. The lecturer has recently experienced surprise, and perhaps their intuitions have had to be revised, or experienced pleasure that their theory has been confirmed. The lecturer is 'in question', and is freshly working on the material as they speak, which has a stimulating effect on their audience (as long as the lecturing is reasonably coherent!). How can you bring this excitement to material that is well rehearsed and elementary?

The most valuable and effective approach is to be genuinely interested in what students are thinking, and how they react to what you say and do. This can actually benefit your research, both by throwing up unsuspected structural problems or wrinkles on well-understood ideas, and by enabling you to re-experience the struggles you have had with your own mathematical ideas. By noting how you help your students with their difficulties, you can perhaps apply the same principles to your own research (for example by recognising characteristic locks and barriers, and choosing to act differently so as to get around them).

Another approach is to seek out other supporting texts for the course and to consider what is different between that author's sequencing of examples, theorems and proofs, and your own. Alternatively, you could look up examination questions from other institutions and see what is similar and what is different about them. You could also find out something about the history of the result, the context in which it arose, and the activities of the mathematicians involved in elucidating the ideas (see Chapter 6).

Most theorems contain either a surprise or a confirmation. Try to locate the surprise signalled by a theorem (for example that not all series can be re-arranged and then summed. or that the scalar product of two vectors is independent of the coordinate system), or the confirmation of intuition (for example that all bases have the same cardinality). If students do not have the underlying intuitions brought out for them, then the importance and effect of the theorem may pass them by. Identifying the intuitions that are either surprisingly contradicted or satisfactorily confirmed may also provide a possible entry for students so that the result becomes meaningful for them. (See *Issue: Locating Surprise*, p181.)

One way to help students learn mathematics is to develop a sensitivity to their difficulties. You may have struggled with some ideas as an undergraduate yourself, or you may have floated through without much effort. In either case, being expert in the topics you are teaching means that it can often be difficult to imagine what it must be like to be a student encountering these ideas for the first time. Consequently, it is not always easy to be sensitive to the difficulties that students may have, and hence not easy to find ways of presenting the ideas and structure in a way that engages students and helps them surmount or circumvent those difficulties. It can be useful to build up files associated with different topics that you teach in which you make notes about interesting things that students say that reveal misconceptions or unwarranted assumptions, particularly apt and effective examples which you come across, and historical information which you notice. (See *Tactic: Catching 'it'*, p95, and *Framework: Concept Images*, p190.)

Tactic: Interrogating Your Own (Analogous) Experience

Tactic:	Rather than trying to re-experience a struggle that has long since been resolved, it can be invaluable to look for analogies between your students' experience of the topic and your own research endeavours, or your own experience trying to learn some new topic for yourself.

The idea is that where students are encountering new notation for new ideas (such as in linear algebra), or a collection of theorems that formalise intuitions (such as in analysis), you can appreciate their difficulties by reflecting upon your own efforts to make sense of new ideas, notation, or complex theorems in reading research papers or in proving new theorems. By pausing every so often and asking yourself what would help you at that moment, or what it was that caused the most difficulty, you can develop an awareness of the life of a student. This in turn can inform the way you structure your teaching.

The notion of seeking an analogy with your own experience is based on two conjectures. Firstly, each technical term in mathematics signals a shift in perspective, a shift in what is stressed and what therefore is ignored, that mathematicians underwent when the topic was being resolved for the first time. Secondly, many – if not most – theorems represent the expression of a surprise at a contradiction to intuitive expectation, or a confirmation of intuition. To gain facility with a technical term it is necessary for students to experience a corresponding shift in their perspective (see *Theme: Shift of Attention*, p186).

Reflecting upon your own thinking can have three benefits:

- it can increase your sensitivity to your students' difficulties;
- it can provide a focus of interest and questioning for you;
- it can inform your own research by sharpening your sensitivity to the strategies which you find useful when doing your own research.

Keeping Fresh

If you have taught the same course a number of times, it is tempting to resort to your old notes rather than prepare fully each time. Halmos (1985, p69) noticed this: 'Another way I keep active is by changing the notation; if there is nothing else I can do, I can at least change (improve?) the choice of letters.'

A change of notation is useful in a number of ways. Students tend to become habituated to a particular notation and can have a hard time recognising a technique or theorem if the notation is different. By changing notation one can try to stimulate students to become more flexible, and to probe beneath the notation to the ideas below. To see the effect for yourself, try the following exercises in your head.

Task: Changing Notation

Say out loud, without making notes, what the roots of the quadratic equation $bc^2 + xc + a = 0$ are as an equation in c.

What is the following defining? For all *f* there exists an *x* such that for all ε such that $|\varepsilon - \varepsilon_0| < x$, $|\delta(\varepsilon) - \delta(\varepsilon)| < f$. *How did you recognise it?*

Now, no-one in their right mind would scramble letters like that! However, it does provide a taste of what it is like for students to meet unfamiliar notation and new ideas, except that you have the advantage of recognising the shape of the definitions even though the letters are mixed up. If you struggled with either of them, even a little, then it may remind you how much even experts depend on a traditional or conventional use of notation to recall basic ideas. Yet it is this very reliance on particular notation that prevents students from being flexible enough to differentiate with respect to *h* or *k*, or to recognise familiar patterns arising in different contexts, such as factoring differences of squares or using integration by parts.

Halmos took a strong position on this issue.

'Changing the notation is an attention-focusing device, like taking notes during lectures, but it's something else too. It tends to show up the differences in approaches of different authors, and it can therefore serve to point to something of mathematical depth that the more complacent reader would just nod at.'

(Halmos, 1985, p70)

See also *Tactic: Multiple Notations*, p13.

Tactic: Varying Notation

Tactic: Try varying your notation, preferably to improve it, but also to keep yourself and your students awake and flexible.

If you have to repeat a lecture to a different group of students, how do you make the second one fresh? How do you avoid expecting them to know what you have recently told the first group?

In addition to the suggestions made above, there are several possibilities, including: constructing new examples to replace the ones already used; experimenting with a different ordering of the material; or using a different pedagogic sequence of example, definition, theorem, and proof. Another option is to work extra hard at being aware of what you are stressing internally, and how you that is manifested externally in gesture, posture, voice tones, or pauses. (See *Tactic: Making an Example be Exemplary*, p29).

Handouts

There are a variety of choices to make regarding handouts. Whatever decision you use (perhaps the students could contribute their opinions), it is useful to have an articulated justification for what you choose to do, and then to act so as to maximise the possibility that students will react in the way you anticipate. Possibilities used by different lecturers include variations of the following strategies.

- Hand out complete notes in advance or at the end of a session or a topic (what are the differences for students?)
- Hand out notes giving definitions, theorems and examples, but leaving spaces for proofs, or just for additional comments as the lecturer talks about her own sense and appreciation of the topic.
- Hand out brief summaries of the most important definitions, theorems, or techniques.
- Make no handouts, but perhaps give references to pages in standard texts.

What use can/do students make of each type? What assistance might they require to make the most effective use of the notes?

What you provide for your students in the way of handouts depends on your aims and intentions. Most people have a mixture of aims, and so use different approaches at different times. It is important that students know both what to expect and what is expected of them!

Aims	*Expected Actions by Students*	*Styles of Handouts*
Cover (explain, teach, transmit, or convey) the definitions, theorems, proofs, and techniques	Study (not just read) notes mathematically; work mathematically on set exercises	Complete notes as if in a book (available in advance or after the lecture) Definitions, theorems, and sample worked examples
Inspire	Appreciate overall flavour; pick up details from carefully working on notes and perhaps texts, not just working through exercises	Notes with headings but details left as spaces for students to fill in as you work through the exposition Explicit references to standard texts Extra suggestions not mentioned in the lecture
Work at understanding, making connections between topics or theorems	Re-construct topics for themselves from lecture notes and text, and, increasingly, independently from text alone	A succinct mathematical summary, perhaps with worked examples or challenging questions to explore
Teach how to carry out required techniques and solve sample problems	Work on 'worked' examples; justifications and theory found in text	Elaborated worked examples displaying choices, wrinkles, and use of theorems

Handouts are sometimes given out at the beginning of a term or course (acting partly as a substitute text), week by week (in advance of lectures), or lecture by lecture (in advance or afterwards). They may consist of photocopies of transparencies or of the lecturers' notes, or be specifically designed. The important thing is to decide on what you expect your students to be doing, and then work to ensure that that is what they do.

Getting students to make their own notes is fine for students who can keep up to your pace *and* attend to what you are saying. However, writing everything on a board means that some students inevitably fall behind, with the effect that they miss the relevance of what you are saying and end up merely transcribing. On the other hand, making no notes during a lecture but working at helping them to understand what you are saying may give students the impression that they are learning, but gives them little to fall back upon. Handing out complete notes gives the impression that students may not need to attend lectures: students need to be clear about what role the lectures play, and of course lectures can play several roles. Some people use as their weekly diet the following sequence: motivational-inspirational; work on some main ideas; go through worked examples.

'Mixed Ability'

The inverted commas indicate that the notion of 'ability' is a construct that is not always helpful: students sometimes display one behaviour one day and another the next. Scoring highly on some tests is not always carried through into examinations, nor into later success as a mathematician. All classes are 'mixed ability', since what is said is picked up on by some but not all students, and individual students respond on some occasions and not on others.

The issue is how to cope with a spread of confidence, sophistication, and technical competence. If you always direct all your attention to those who are least confident and who need details and time, you may bore the quicker students, or those who have more experience or sophistication. On the other hand, if you always direct your attention to the quicker students or to those who readily call out answers or at least fix their attention on you, you may lose the others. The best policy is a mixture, but it is hard not to fall into patterns at one or other extreme.

Keeping quick-thinking students occupied is not very difficult: students who grasp ideas quickly can be encouraged to generalise, vary, and extend ideas. It can become a standard practice that students who wish to are invited to generalise and extend, while others are occupied in constructing and checking examples for themselves. Students who are not yet experienced in mathematics and are shown *how* to study mathematics are more likely to gain the requisite sophistication quickly. Encouraging them to reconstruct ideas, theorems, proofs, or techniques through specialising and re-generalising can do this. (See *Tactics: Specialising, Generalising and Counter-examples,* p10; *Boundary Examples,* p14, 136; *Expressing Generality,* p14; *Say What You See,* p20, 30, 91. See *Issue: Time – Coverage and Pace,* p164.)

Encouraging Exploration and Mathematical Thinking

If it were sufficient to tell students succinctly but clearly the meaning of definitions, the content of theorems, and how techniques are applied, then mathematics teaching would have been perfected long ago, for there are many textbooks which claim to do this. However, there is more to learning than reading exposition and doing exercises, and there is more to teaching than laying out the ideas and methods clearly. Learning is about taking the initiative to reconstruct ideas for oneself under the guidance of a teacher, and teaching is about providing appropriate guidance.

There is never enough time for students to explore around the topics they encounter, but if they are never encouraged to explore, or pointed in useful directions, their impression of mathematics may be of a body of established truths which simply have to be memorised. This applies particularly to students for whom mathematics is a service subject, a painful means to some other end. If you do not encourage them, they may never experience the pleasure and pain of trying to sort out an idea through many failed attempts, searching for examples, trying to generalise and then trying to capture and write down a method or an argument. They may remain stuck in inefficient study habits that do not enable them to learn effectively.

Students can be directed to websites which describe applications and models that employ the ideas being discussed, and they can be asked to summarise one or more case studies for their colleagues. It can also be worthwhile to direct them to websites that contain animations and interactive worksheets; these can include worksheets that you develop specifically for them, or tape-frames (see *Tactic: Tape-frames*, p68). You can post a problem of the week, and offer challenges during lectures. However, offering too many things is likely to overwhelm students, who may then do nothing. On the other hand, offering nothing gives the impression that there is nothing else to mathematics other than pre-digested theorems, proofs and techniques. If you take an evident interest in what students do, such as mentioning a challenge or a student's attempt at it during a lecture, they are more likely to take up challenges than if they are simply seen as extra work.

Tactic: Extending Core Material

Tactic:	Suggest variants on standard tasks that are liable to bring the students up against extensions or more sophisticated ideas, perhaps by turning doing into undoing (see *Theme: Doing and Undoing*, p193). See Appendix A for examples.
Tactic:	Offer exploratory questions designed to bring students up against the main ideas of a topic. (Again, see Appendix A for examples.)
Tactic:	Invite students to model familiar situations using the current topic, or to locate where the current topic is being used within a model presented to them.

A classic example is provided by the Millennium Bridge in London, which had to be closed on its first day because the unexpected number of people made it sway in ways that had not been anticipated. Government policies can often be modelled and critiqued very simply using ideas from a course.

Tactic: Ask students to construct 'boundary examples', where a change in any one of the conditions will turn it into a non-example, or from a non-example into an example (see *Tactic: Boundary Examples*, p14, 136).

See also *Tactics: Student Generated Exercises*, p16; *Expressing Generality*, p14; *Particular – Peculiar – General*, p88.

Tactic: Theme Exploitation

Tactic: Ask students to suggest how a given theorem or topic exploits or illustrates the mathematical themes discussed in this book, such as doing and undoing, invariance amid change, or freedom and constraint. (See *Themes*, p192.)

Providing Additional Support

There is no reason to expect that all students will receive sufficient support from simply attending lectures and tutorials, writing up their notes, and doing exercises. Students taking mathematics because they *have* to, as a necessary requirement of the subject that is their real interest, are particularly likely to need more support but be reluctant to devote time to obtaining that support.

Various structures can be used to help support students. Some institutions have a drop-in centre where students can get help with assignments and with studying techniques. (It is important to provide help with both of these, as providing immediate assistance does little for the student unless there is also some longer-term assistance in how to study and in problem-solving techniques.) Some institutions set up informal networks of students in one year who help students in the year below them. This is good for all concerned, as the more experienced students are seen to struggle with ideas they are meant to have met, and learn from trying to explain things to others.

Tactic: Student Mentoring

Tactic: Assign students with more experience to be mentors for less experienced students (say second years mentoring first years, and perhaps third years mentoring second years). Suggest a weekly informal meeting to discuss problems and to get help with problems, or have mentors specify consultation times when they will be available.

Some students have reported that helping other students made a significant contribution to their own learning. Having to explain something to others can be of great help to a mentor, as it stimulates them to reconstruct their own knowledge, as well as being helpful to the student seeking help.

Tactic: Tape-frames

Some mathematical topics are well known to cause most students some difficulty, and many students a great deal of difficulty. For example, the definition and meaning of functions, limits, cosets, and uniform continuity are known to be major hurdles for students. So are choosing which method to use in integration, which probability distribution to use as a model, where to locate the forces acting in a given situation, and how to find a particular solution of a non-homogeneous system of differential equations.

Tactic: One approach is to develop a brief exposition on audiotape with supporting notes, symbolic expressions, and diagrams, and lodge it in the library, available for students to use. You could recommend that they work through a specific one prior to the topic being mentioned in your lectures, so that they are prepared for it, and have some background to your discussion. A first draft can be prepared by making some notes of examples or definitions, talking through them on tape, and then duplicating the notes. These can then be refined over a period of years as a result of student comments.

A more detailed structure can be used, such as the tape-frame, developed at the Open University. It consists of a series of printed frames, each with a heading and a number, and an accompanying audiotape. This enables the tutor to say things that they might be embarrassed to print (for example how they informally think about the topic), while at the same time giving the student the effect of a tutor talking in their head. The frames are usually a mixture of headings that summarise more detailed comments on the tape, worked examples, and tasks for students to try which mirror the worked examples. Tape-frames provide the lecturer with an opportunity to refine and hone their exposition of basic but troublesome ideas, and then to focus on wrinkles, applications, and more subtle matters in their lectures. Once made, tape-frames can be stored in the library and used by many students year after year.

Tactic: Computer-based Mathematics

Computer- and web-based versions of tape-frames are increasingly used by many institutions. These permit animated diagrams and active calculations, so that students can modify or adjust parameters for themselves. In order to make best use of these resources, students need to know how to work on the resources. There is a tendency for multimedia to feed addiction to stimulation: users are eager to get on to the next page in case it is even more exciting than the current one, with the end result that they see a great deal, but not much sticks. In comparison with tape-frames, there is also a loss of some of the personal intensity of the earphones, in addition to the temptation to tinker with the keyboard and mouse rather than work on the ideas.

Tactic: If you are familiar with a symbol processor such as Mathcad, Mathematica, or Maple, then you might want to construct worksheets with embedded formulae which students can modify and use to explore the topic.

There is a strong temptation to simply reproduce one's usual exposition with the addition of moving pictures and executed calculations, rather than to help students to take the initiative and make sense of the topic for themselves using what is available. These resources seem to be most effective when students are enabled to construct their own experiments and undertake their own explorations, perhaps with some assistance in the form of procedures provided in advance.

Be warned, however: a tape-with-notes is relatively easy to construct, but a tape-frame is a more stylised version, and takes a bit longer. Symbolic processor worksheets take even longer to develop and to make them robust and helpful. Animations also take a lot of work if they are to animate what is dynamic while drawing attention to what is invariant, without misleading students. Any of these forms can and should be tested by students, and modified on the basis of their feedback.

Reflection

Task: Reflection

Make a list for yourself of at least six of the tactics proposed in this section. Select two that you might be able to use soon, and prepare yourself to use them by writing down what you think you would say and do, and then imagining yourself actually using it.

Comment: Imagining yourself doing something gives you the opportunity to anticipate how students might respond, and how you might cope with that response. Do not forget that you can always 'cut and run' back to your habitual practices, and that you can ask students for their co-operation as you explore with them ways of improving your teaching and their learning.

3 Tutoring

'The hardest part of teaching by challenging
is to keep your mouth shut, to hold back.
Don't say, *ask!*
Don't replace the wrong A *by the right* B,
but ask, "where did A *come from?".*
Keep asking "Is that right?, Are you sure?"
Don't say "no"; ask "why?".'

(Halmos, 1985, p272)

Introduction

In lectures, students are invited to enter the lecturer's world, guided by their experience and familiarity with the territory. In tutorials, the tutor can take the opportunity to try to enter the students' world. The aim is to see what your students are seeing, in the way that they are seeing it. You can then hope, as a result, to be able to make useful suggestions from which they can learn.

The most common trap for any tutor lies in the shift from trying to be sensitive to their students' difficulties, to trying to sort out those difficulties. Once you work out what you think has been misunderstood, it is tempting to return to lecturing mode, telling your students the answer, or what they should be thinking. Sometimes this works, but often it does not, as showing students your solution to a problem is an attempt to drag them into your world. If your students have genuinely struggled with the problems first, and if, in your presentation, you indicate the choices you made at each step, then students may gain a great deal. However, it is worth remembering that, once you move back into expository mode, students are again 'being shown'. In order to learn from being shown, students have to be able to make sense of what they are shown, and for this they need to know how you thought, not just what you did.

Tutorials provide an opportunity to work at guiding students in the development of their mathematical thinking, rather than simply doing the thinking for them. They provide time to encourage students to struggle publicly as part of the process of sorting ideas out. The old adage 'you really learn when you are called upon to teach' suggests that a fruitful support for learning is to get students to teach each other. As a step towards this, getting students to explain their thinking, even where they know it is incomplete, can help them to sort out what they are confident about and what they are still unclear on.

Tutorials can also provide a time for students to ask questions about ideas, examples, theorems and techniques that are bothering them. The trouble is that very often students do not take that opportunity. They do not ask questions at all or, if they do, they are of the, 'Can you do the third problem please?' variety. It takes time and effort to develop student confidence and expertise in asking questions. At first they usually do not want to reveal how little they understand, or they do not know how to ask questions other than, 'Can you go through ... again please?'.

Confidence can be built by developing a conjecturing atmosphere, where students feel free to air their conjectures without fear of ridicule.

Since many of the tactics proposed here are related to or are based on the same principles as tactics already suggested, there is a certain amount of repetition in this chapter.

Conjecturing Atmosphere

Getting students to struggle publicly to describe what they are thinking, to 'say what they see', and to reveal where they do not see or what they are confused about, provides access to their thinking and enables you to help them. However, it requires a supportive, 'conjecturing' atmosphere in which everything said is understood to be open to reconsideration and modification (in contrast to an atmosphere in which one only speaks when one is confident one is correct). It is essential that students trust that no one will make them look foolish. In a conjecturing atmosphere one does not say, 'No, you're wrong', but rather, 'I invite you to modify your conjecture', 'What about this example?', or some other more focused question. Similarly, the act of modifying a conjecture can be seen as a sign of progress.

In a conjecturing atmosphere, people still choose to say things even when they know that they are unsure about some detail, or know that they may not be able to complete the sentence. They say things with the expectation and intention of modifying what they have said. Those who are confident often choose to listen to others and then to offer examples and counter-examples. Every contribution is taken seriously. Where someone disagrees, they offer a counter conjecture or a potential counter example.

It is vital, when ideas are tumbling around in your head, to externalise them, to express them out loud or on paper so that they can be pinned down and examined. Much of the confusion in mathematics arises simply because ideas are never examined, and never tested against examples. Once they become established, they are hard to shift. This is why it helps, when working on problems (from a class or in research), to write down (and label) conjectures, and then to focus on those.

Tactic: Assessing Degrees of Confidence

Tactic:	If a student asserts something loudly and apparently confidently, try asking how much they would stake on their claim. Alternatively, use the word *conjecture* in relation to their assertion, and then ask others to see if they can find a counter-example.
Tactic:	On assignments, ask students to rate their confidence in their solution on a scale of 1 to 5. The purpose is to get them to think about what they have done in addition to striving to get an answer. It may also be of help to you when you are marking.

Tactic: Voting on Conjectures

Tactic: To generate conjectures, propose a problem to which students are likely to offer different answers. Then tell them that they will have to vote on the conjectures, so they must make sure that the one they favour is on the list. Actually carry through the vote as well. The aim is to get students to test their intuition by choosing one option, rather than waiting to be told which is correct.

Committing oneself, however temporarily, is part of taking the initiative. Developing ways of working in which a conjecture is made and then tested or challenged is much more useful than reinforcing a way of working in which your students wait to be told what to do and how to do it. Once a vote is taken, get people to try to justify their choice or to offer examples that contradict the others. In this way you can demonstrate how practising mathematicians approach conjectures.

Scientific Debate

Marc Legrand (1999) emphasised the importance of immersing students in a strongly mathematical culture, using what he called *scientific debate.* He saw students as participants in a scientific community whose methods of development include making conjectures and then modifications, proofs and refutations (Lakatos, 1976). The essence of a scientific debate, and of a conjecturing atmosphere, is that people are eager to try ideas out and neither embarrassed nor ashamed to make a mistake. Value is placed on modifying one's conjecture to take account of criticism, not simply on getting correct answers.

The intention is to create the conditions under which fruitful mathematics is done at any level of sophistication. If students' only experience of mathematics is from honed texts, prepared lectures, and homework exercises, then they are likely to miss the excitement and the essence of mathematical thinking, and this may seriously hamper their intellectual development. Consequently, Legrand has worked at how to engage students in mathematical discussion.

Guiding principles for scientific debate include the following.

- *Disturbance:* students must face up to and deal with any conflict with their current ideas (the need for disturbance to produce conditions for learning can be found in the work of most educational thinkers since Socrates).
- *Inclusiveness:* every human being should have the opportunity to understand the deep meanings of what we are striving to teach.
- *Collectivity:* collective resolution of issues shows how to work with contradictions and to respect the views of others.

A scientific debate can arise spontaneously when students query a statement (for example does $f(A \cap B) = f(A) \cap f(B)$?), or it can be stimulated intentionally by asking students to make a conjecture regarding some problematic issue.

The critical point is what the tutor does once conjectures start to arise. Indicating through words, tones, gestures, or posture which conjectures you agree with and which you do not will quickly turn the session into 'guess what is in the tutor's mind'. Consequently, it is vital to adopt a neutral stance with regard to contributions, giving everyone a chance to make a contribution and making sure that all contributions are considered and valued. As an expert, the tutor must attend to the process as much as to the content, making sure that no assertion is rejected without counter-examples and none accepted without justification. Not every conjecture has to be 'dealt with' immediately. It is often more useful to store conjectures on a board as they arise, so that they can be returned to, modified and then accepted or rejected later.

You cannot expect students to begin acting immediately in the same way as a group of mathematicians round a board or at coffee, but that is what you want to work towards. You also cannot expect students to be aware of fine points and of novel counter-examples, so, if the group becomes satisfied with an argument that you know has flaws, you will have to be prepared to offer either a counter-example or a question which directs attention to the difficulty. If the group runs out of steam, you can offer an instructive example, or even make a conjecture (not necessarily a correct one) to get them going again. Your aim as a tutor can be to reduce the density of your intervention over time so that your students start to participate in and monitor the scientific debate themselves. (See *Tactic: Directed –Prompted – Spontaneous*, p91.)

The format and aims of a scientific debate can be adjusted to promote other forms of discussion among students and between tutor and student, when this is appropriate.

Tactic: Generating Discussion

Tactic: Get several students to present their solutions to a problem on the board, and then guide and chair a mathematical discussion of those solutions. If a silence develops, keep calm and concentrate on the mathematics, rather than being caught up in the fact of the silence. Bring up simple questions such as: 'Can anyone provide an example (or counter-example)?'; 'Can someone restate the difficulty in their own words?'; 'Can someone draw a diagram to illustrate what is going on?'; 'Does anyone know a theorem that uses that idea?'.

Tactic: Get a student to explain the question as posed (in their own words) to others, as the beginning of a debate among the students about what the problem really is. Alternatively, get a student just to read the question out loud; this often reveals where the uncertainties lie.

Tactic: Stop every five minutes or so during a session and pose a question to students. If you receive several different responses, get them to discuss them in pairs and decide which is correct (see *Tactic: Talking in Pairs*, p49, where a brief reference is made to the possibility of doing this electronically).

Use the tactics *Talking in Pairs*, *Particular – Peculiar – General*, p88, *Scrambled Proof*, p94, *Advising Students How to Make the Most of a Worked Example*, p97, or indeed any of the suggestions in Hubbard (1995), Mason *et al.* (1982), Watson and Mason (1998) or Burn, Appleby and Maher (1998).

These tactics are all designed to provide students with experience of tackling the unknown, whether assignments or text, in the hope that they will begin to use such tactics for themselves when they are studying at home.

Asking Students Questions

Asking students questions seems relatively straightforward, until you find that they sometimes remain quiet, waiting for an easier one or one that they understand. There is an art to asking questions that stimulate students to take the initiative. The most important thing is to build an atmosphere of trust so that, when they do not understand, they nevertheless feel empowered and encouraged to ask. As long as students are sitting waiting for something to happen, the initiative lies with the tutor, and learning is stymied.

The most effective questions are ones which are genuine requests for information and expressions of interest, either asking students what they did next or where they got stuck, without implying, by tone of voice or otherwise, that they should have been thinking or acting differently. Questions which prompt students to act are especially useful, for instance asking them to modify an example so that it has additional or fewer specific properties (for example asking for functions which are differentiable everywhere except at a specified number of points, or except at specified points; see *Tactic: Particular – Peculiar – General*, p88). Questions that are typical of the sort that mathematicians (or experts in the field of interest to your students) are the most effective, for they show students how experts enquire. (See *Forms in Mathematics*, p124, for examples within mathematics.)

Funnelling and Guessing What is in Teacher's Mind

If students will not ask questions, then it is natural for the tutor to start asking questions instead, but there are many pitfalls here, too. When asked a question, students expect the teacher to be looking for a specific answer, so they either play along by trying to work out what that might be, to the extent of guessing, or they rebel. Either way they are not learning much.

When students rebel against these kinds of questions, usually by remaining quiet, the tutor can be drawn into a sequence of ever-increasingly specific and explicit questions, searching for something that the student can actually answer. This process was referred to as *funnelling* by Henrich Bauersfeld (1994). Funnelling is played out in classrooms all over the world every day, but it is not, on the whole, very successful. Even if the students are genuinely unable to answer, the 'blisteringly clear logic' (from the tutor's point of view) is likely to be an opaque mystery to them. Consequently, even if a group of questions is located which students can indeed answer, there may be little learning achieved: being led through a sequence of questions which they can answer is unlikely to sort out what they do not know.

There are alternatives, however, as the following tactic demonstrates.

Tactic: Anti-funnelling

Tactic: When you catch yourself funnelling, try stopping and being explicit: 'I am trying to locate the point of difficulty. Can you help me?', or even, 'If you read the problem (page of text, or whatever) out loud to me, I might be able to find out where the difficulties lie.'.

If you catch yourself playing 'guess what is in my mind', then stop and be explicit: 'I am beginning to realise that I had a particular answer in mind.' Then you can decide either to simply tell the student, or to adjust your approach.

Getting Students to Ask Questions

If only students would ask questions like, 'Can I try and say that, please, because I don't think I can?' in response to something a tutor says, or, 'Working on this problem I can get so far, and then I get stuck', or even, 'I don't know where to start on this problem or on problems like it', a tutor's life would be easy. A student who is willing to ask these sorts of questions is relatively easy to help. It is the students who do *not* ask, who do not know what to ask, who probably need the most help.

Students entering college have a long history of being asked questions in school. Unfortunately, it is very likely that those questions will have been teacherly questions: questions to which the teacher already knows the answer. Questioning in schools is often used for purposes of control, whether testing what students remember, or trying to keep their attention on the topic at hand. Thus students may not be used to being asked questions to which the questioner genuinely does not know the answer. Furthermore, not having been in the presence of genuine enquiry, students may not be in the habit of asking real questions themselves. They may feel they are limited to questions of the form, 'Can you tell me how to do this?'.

A major frustration for teachers of mathematics is that students do not naturally seem to want to know why things work. For example, Roberto Baldino reported in an email discussion that when he carefully developed the principle of L'Hôpital's rule, students were uninterested in his 'why' questions. When he asked them why it worked, they responded with 'What do you mean? It works because this is the way you do it.' By contrast, when he presented the technique itself without reasons, he found students asking him why it worked (see also *Tactic: Locating Surprise*, p181). It may be that when a lecturer presents reasons and proofs, their students accept them even when they do not understand them. Their sense of 'why' may then simply be 'because it works', backed up by 'an expert claims to know why'. This suggests that you conduct an experiment: every so often, offer a technique that has some surprising element in it, without explaining why it works, and see if your students ask for justification. (See also *Tactics: Interrogating Your Own (Analogous) Experience*, p62, and *Issue: Locating Surprise*, p181.)

The switch from justification based on 'because it works', or 'the text says it works', to appreciating *why* something works based on previously understood ideas is not an easy one, especially for students for whom mathematics is at best a tool and at worst a hurdle. One reason for this difficulty in appreciating the essence of mathematics is that students often have a long history of being told to accept things without questioning them. They often have neither the experience nor the structures within which to ask such questions, and they have no idea of what an answer should look like. If it works, then that is all that matters. The notion of deduction from first principles is very sophisticated, and made even more tenuous by the strangeness of being taken through a formal justification of things they thought they already knew. (See also *Tactic: Being Explicit About the Enterprise*, p56.)

There are two obvious approaches to overcoming this past experience of students. One is to engage them in mathematical discussion and scientific debate (see *Scientific Debate*, p73; and *Issue: Discussing Mathematics*, p176). The other is to expose them to a broad range of the types of questions which mathematicians ask themselves (see *Forms in Mathematics*, p124). For example, asking students structural questions such as 'How does this example illustrate that theorem?', and 'What theorems does this technique depend on?', is likely to engage them in thinking about the structures being offered. This is the aim of the many tactics concerned with encouraging students to construct their own examples and questions. (See *Tactics: Linking Terms*, p84; *Story telling*, p82*; Advising Students How To Integrate A Topic by Weaving a Story*, p98.)

Another way to stimulate students to ask questions is to get them to read extracts from a text or a problem out loud. It is amazing how much is revealed by where they pause and how their voice tones change. A further way is to invite them to bring along questions based on their experiences that might (or might not) be tractable using the available techniques. For example: a film of a chimney being demolished may show the chimney breaking in two before it hits the ground and raise questions about why it does so; a visit to a Gothic cathedral may raise questions about the role of flying buttresses; a reference to the Mary Rose might raise questions about why it capsized. Further information about such phenomena may be found on the Internet, and similar questions can be found in the *Flying Circus of Physics* by Walker (1975).

Worked Examples

'Affirmative or negative, it's examples, examples, examples that, for me, all mathematics is based on, and I always look for them. I look for them first, when I begin to study, I keep looking for them, and I cherish them all.'

(Halmos, 1985, p64)

A session working through examples is a standard feature of many tutorials, but what is the student to make of it? Are students aware of what is particular to the example and what is general about it? Can they see the general through the particular, or is the example yet more 'text' to be re-read or even 'learned'?

Students are forever asking for more worked examples. If an assignment question is unlike any they have seen before then they may feel cheated. Their expectation is that your job is to teach them how to do whatever the have to do in order to pass (and presumably, as a result, to learn what they need to learn). Glenda Anthony quotes a mathematics student as saying,

'Good worked examples are the most important part of my learning initially, otherwise I've got nothing to base my understanding on.'

(Anthony, 1994, p131)

However, in the seminal study which introduced the notion of a 'hidden curriculum', Snyder observed that students

'... almost always translate [what the faculty says they expect from them] into a series of discrete, more or less manageable tasks which they infer is the actual basis for the grade their professors will give them.'

(Snyder, 1970, p5–6)

The more comprehensive the worked examples provided, the easier it is for students to repeat the format, but the more likely they are to miss the provider's intention, and fail to appreciate the generality. Brousseau captured this in the *contrat didactique*, which can be stated as an endemic *didactic tension.*

'The more clearly the teacher indicates the behaviour sought as indication of understanding, the easier it is for the students to display that behaviour without understanding.'

(Brousseau, 1984, 1997, and see also Laborde, 1989)

(See *Issue: Implications of Student and Teacher Expectations,* p166.)

To meet student wishes, and to be seen to be helpful (for example when quality assurance is being inspected), it is very tempting to provide what the students want: worked examples of every type they will meet in assessment. However, the effect may be that, although they can deal with routine questions, they are at sea when asked to move into new territory.

Some students do not realise that merely to have read, or even worked through, a mass of examples is not usually sufficient. They need to have an appropriate technique come to mind when it is needed, and to do this requires them to have internalised its ideas and purpose, not just the sequence of steps involved.

It follows that, when working through an example for students, more is required of you than simply to carry through a sequence of steps. Students need to know what an expert thinks about as they are working on a problem. They need to know which features attract the expert's attention and why, what choices the expert is making, what the expert sees as particular to the situation, and what they see as generic. Often there are inner monologues that float through an expert's mind as they work. Students need to know about these too. (See *Tactics: Exposing Inner Monologues*, p57; *Inner Moves*, p20.)

It may seem to students that it is less effort to follow lots of examples provided by the lecturer than to attempt to understand what is going on. They may need to be convinced, especially in a service course, that it is actually less effort to appreciate what the topic is about, what the theorems say and what the techniques achieve, than simply to rely on memory. A student who can reconstruct a technique is much more flexible than one who relies on memorising meaningless steps. Taking the initiative themselves to reconstruct ideas, by using examples of mathematical objects and worked examples of the use of techniques and application of theorems, is much more effective. However, students often need persuading of this, particularly in service courses.

The best way to persuade students is to help them to experience what they can do (see *Theme: Mathematical Powers*, p184), and to be made aware of their potential. At first this can be done explicitly, then more and more implicitly (see *Tactic: Directed – Prompted – Spontaneous*, p91; *Framework: Awareness, Motivation, and Behaviour*, p190). It helps if they can see the tutor using these powers explicitly as well, of course.

Probably the most useful worked examples are ones that your students have constructed for themselves, perhaps using others as a basis. Students who can describe to themselves what a 'problem of this type' looks like, or who have generalised from one or more worked examples, are much more likely to be recognise the need for the technique, and be able to reconstruct it when needed. (See *Tactics: Making an Example be Exemplary*, p29; *Student Generated Exercises*, p16; *Particular – Peculiar – General*, p88. See also Michener (1978) for different forms examples can take and their possible roles in teaching.)

At issue, then, is how to make the most of worked examples.

Tactic: Making the Most of a Worked Example or Exercise

When experts are working through an example, they are aware of the general concept being exemplified. When students are watching an example being worked through, however, they usually have little or no idea of the generality being exemplified. This means that they often do not recognise whether, when an action or an object appears, that is what happens generally, or just in this particular case.

Sometimes students will ask a tutor, 'Where did you get that 3 ... ?', indicating that their attention is focused on some detail. Often such a question comes immediately after the tutor has worked through the example. A plausible reaction is to rehearse the computation from which the 3 arises, because that is what the question seems to be asking. However, it may be that the student is actually enquiring about the choice made to use the method that produced the 3, not its result in that particular case. They may even be asking for a conceptual justification for doing the calculation. Students are much more likely to be able to follow a sequence of calculations than to know how to decide what calculation to do, and even less likely to appreciate why that choice is made and why the technique works. It is easier to follow passively (waiting for each new step) than to follow actively (anticipating the next step); it is easier to follow actively than to do it for oneself (see *Assent – Assert*, p82).

Tactic: When you are working through an example in front of your students, or even on paper, try to be aware of what you are emphasising, of what you see as particular to the example and what you see as general. Then include marginal notes that refer explicitly to the choices you are making, to what inspired you to think of doing the next step, and to what features you are stressing (mentally) as you proceed. Where possible, include remarks about why the technique works in this case, and what it is about the particular example that suggested to you that you should use that technique.

Although this sounds obvious, there is usually room to be more explicit, requiring you to become more aware of what you yourself are stressing (and consequently what other things you are ignoring).

For example, when diagonalising a matrix for students, do not just write down the determinant of $A - \lambda I$ in the particular case; rather, write down the general determinant, and indicate that you are writing down in symbols the definition of an eigenvalue. If your aim is to proceed directly to eigenvectors, write down the definition in general and then remark that you are writing down the definition of an eigenvector in symbolic form. Only then write down the specific matrix A and start working through the example. If it seems useful, be explicit about moving from the general to the particular, and carry on this level of commentary throughout. It takes longer, but it may occasionally stimulate you to re-think your approach, and it will certainly benefit your students, even making their work easier to mark if they start adding similar explanations in their assignments (see *Tactic: Being Clear About the Mark Scheme*, p144).

Tactic: A variant tactic is to spend a little time every so often (10 minutes once every two or three sessions perhaps) *technique bashing*. This involves working through a problem yourself out loud, articulating as much as you can of your inner reasoning as you carry out the technique. Encourage students not to take notes, but to try to enter into the board or screen (rather than merely looking at it) so as to participate mentally with you. (See *Tactics: Inner Moves*, p20; *Exposing Inner Monologues*, p57.)

For example, integration by inspection involves looking at the function to be integrated as a composite function and then experimenting. Be explicit about how you do this when you are doing it in front of students.

Tactic: Ask your students (and yourself) what is generic and what is particular about a particular example. Spend time explicitly describing what it is about the example that makes it exemplary. This is good practice for axiom or definition checking, and checking that the conditions of a theorem are satisfied.

Over time, reduce the explicitness with which you direct their attention to making good use of examples. As your students gain confidence, it is better if they can be weaned off such support. (See *Tactic: Directed – Prompted – Spontaneous*, p91).

Tactic: Ask students which features in a given example can be changed while still allowing it to remain an example, and what features, if changed, would stop it from being an example.

For example, what features guarantee that a sequence converges to 0? Starting from a given sequence, alter it in some peculiar way that preserves convergence (make the one millionth term a 7, for example), and then in some fashion that can be generalised (so that it is seen as a generic modification). Repeat the process, but with the modifications making the sequence fail to converge. (See *Tactics: Particular – Peculiar – General*, p88; *Boundary Examples*, p14, 136; and see *Issue: What is Exemplary About an Example?*, p173)

Generic Examples

Courant quotes Hilbert as saying that

'If you want to solve a problem, first strip the problem of everything that is not essential. Simplify it, specialize it as much as you can without sacrificing its core. Thus it becomes simple, as simple as it can be made, without losing its punch, and then you solve it. The generalization is a triviality which you don't have to pay much attention to.'

(Courant, 1981, p161)

This is the essence of how many mathematicians work: they find an example that illustrates what can happen in a particular situation. The particularities of the example can make it easier to think about, and provide the basis for moving from the particular to the general. They then look for the general principles they used to construct the solution, and attempt to adapt them to the solution of the larger problem. In many cases, simply replacing the particular parameter values used by generic variables allows the proof to go through otherwise unchanged, making the generalisation indeed trivial. The only obstacle to generalisation arises if properties specific to the particular values used form an essential part of the solution. When they need to reconstruct the idea through lack of recent use, they return to the example and literally reconstruct the proof, again looking through the particular that reminds them of their previous work, to see the general. Halmos addressed this as follows.

'Where can we find examples, non-examples, and counter-examples? Answer: the same place where we find the definitions, theorems, proofs and all other aspects of mathematics – in the works of those who came before us, and in our own thoughts. ... we find them first, foremost, and above all, in ourselves, by creative thinking.'

(Halmos, 1985, p64)

Students could usefully undertake the same activity: working on an example to see through the particular to the general, informed by the theorems, proofs and worked examples provided. (See *Tactics: Student Generated Exercises*, p16, *Specialising, Generalising and Counter-examples*, p10.)

Assent – Assert

There is an enormous difference between *assenting* and *asserting*. Students can sit at the back of the class and say to themselves 'I knew that', 'I could have done that', when we, as tutors, know it not to be true, or at least unlikely. Similarly, students can read a text or listen to someone and believe that they have understood, but all they are doing is assenting to what has been said. It is a cliché, but none the less valid, to observe that you only really learn something when you try to explain it to someone else (asserting). Until students take the initiative, until they try to assert what they think, they often cannot locate their gaps and weak-points.

It is a hallmark of experts that they can use technical terms with ease to express what they are thinking. By contrast, novices stumble over technical terms, and often use them inappropriately. Many seem to prefer to avoid them if possible. However, when students struggle to use a particular word, or break down in the middle of describing how a technique works, the tutor gets evidence of where the student needs assistance, and may be able to help very quickly and efficiently. It is vital to establish an atmosphere in which students are eager to assert what they think when they are unsure rather than only when they are sure, to try ideas out with the intention of modifying them (see *Issue: Establishing an Atmosphere of Mutual Support*, p144).

The central issue for tutorials is how to get students to take the initiative, to do more than merely assent to what the lecturer has said, or to what the tutor does or says. One approach is to reduce how much you say and do, and to stimulate students into taking the lead. However, this can prove to be very frustrating for both tutor and students, especially if the students have established habits of assenting rather than asserting.

A class of tactics have arisen in work with students over the years that go some way to addressing these difficulties by provoking students to assert rather than merely assent.

Tactic: Story Telling

Tactic: After a period of work on some structured exercises designed to lead students through the use and meaning of a technique, ask them to call out some of the technical terms used in the topic, and, if necessary, augment the list yourself. Then ask your students to use those terms in a sentence or two in such a way as to show how the terms are connected, or how they are used in the topic. Students are very likely to struggle to make sentences, despite having read similar sentences and used them over and over again during the preceding work. The following sentence does not count: 'In this topic we used the terms …'!

For example, in discussing change of basis, terms such as the following might be expected.

transition matrix	standard basis	codomain
new basis	linear transformation	representation
domain	matrix	

Students can be asked to write their sentences down, or to tell them to a friend. The friend can then comment on the sentences, and together they can refine them. If they get stuck, they should have recourse to their tutor, and the sentences could be handed in or written on the board. Once students become more confident mathematically, they could be encouraged to try them out to the whole class for public discussion and modification. Hearing other students' versions can help students clarify their own in ways that hearing the tutor's version cannot.

The technical terms can also be put on cards and used as a study device: each student can select some cards and try to link them all together in a meaningful sentence. Where they find themselves struggling, they can ask questions like, 'How might I complete this sentence ...', which is much more attractive to respond to than the usual 'Could you explain ... again please?'.

Tactic: Arranging and Sorting

Tactic: Get students to prepare a set of cards, each with one technical term or the name of a theorem on it. Get them to arrange the cards in a way which indicates connections between the ideas.

Get students to prepare cards each with one problem or exercise on it, perhaps taken from the ends of sections in a text. Ask them (individually or in pairs) to sort their cards in any way that they think is appropriate. Then get them to go around and look at other people's methods of sorting, and attempt to write down the criteria they think were used to carry out the sorting.

Variant: Make a collection of mathematical terms, taken from a topic such as linear algebra. Write down some specific mathematical objects which are examples of these different terms, and write down some statements, some true, some false, which are relevant to each term but which do not actually employ those terms. Now ask students to arrange them in some way that they feel makes sense, as above.

For example, consider a list of terms such as linearly independent, linearly dependent, basis, and dimension. Now offer objects such as $\{[1,2,3],[2,4,6]\}$, $\{[1,1,0],[0,1,1],[1,0,1]\}$, or $\{[a,2a,3a] : a \text{ real}\}$. Finally, add some statements such as 'this set is linearly dependent', 'this set is linearly independent', or 'this set spans a vector subspace of dimension 1'. Then ask students to arrange the three collections in ways that make sense to them.

It is amazing how many different arrangements will arise. Also, seeing that other people sort differently is useful in itself, for it opens the possibility of thinking differently. Trying to work out what criteria were used by someone else is also a good way to become aware of the way you are looking at things, and to recognise that there may be more effective ways. The tutor can also include their own arrangement as one to be considered, they wish. It is also helpful to see that different people can arrive at the same method of sorting using different criteria.

Tactic: Linking Terms

Tactic: Get students to prepare a set of cards, each with one technical term, or the name of a theorem or technique, on it. (Students can submit words, or you can provide them). Now get two students to each choose one card at random, and invite anyone to construct a sensible sequence of sentences that starts from one word and ends using the other, making coherent mathematical sense in between. To work well, this activity requires students who are sufficiently confident to try ideas out even if they get stuck.

For example, given *fraction* and *integral*, it was easy to produce, 'In order to find the integral of a rational polynomial you usually need to use partial fractions.' Given *quotient group* and *linear independence*, it was a bit harder to find, 'Quotient groups generalise modular arithmetic. A special case of modular arithmetic occurs when you have a prime modulus, and the result is a field-structure. You can construct vector spaces over fields, and so have vectors which are linearly independent modulo a prime." The effort was rewarding, however, for students were exposed to each other's ideas, and were helped to see links they had not seen before.

The aim is not to have really clever sentences, but to stimulate students to link ideas together. I found that allowing silences, without intervening to break them, encouraged people to come up with more examples, and that people started adding things like 'I've always wondered about ...'. These could then be discussed, or not, by the group. It may be relevant that the time I used this tactic to greatest effect was late at night during a concentrated one-week summer school devoted to the mathematics of linear algebra, group theory, geometry and analysis.

Tactic: Scrambled Correspondences

Tactic: Give students a scrambled collection of technical terms, diagrams illustrating those terms, and symbol-based examples of the terms. Ask students to decide which items belong together.

Variant: Provide an incomplete collection and have students add the missing elements.

Angelo and Cross (p126–131, and p148–153) suggest something similar, but recommend that the tutor should carry out the task themselves beforehand, which is wise advice for any tactic: at least do it yourself when you ask students to do so, and preferably check it out in advance!

Some tutors have expressed concern that in exercises such as the ones described above, students might not achieve a correct statement. They might even leave the room with a partial or incorrect articulation of some mathematical ideas. Should a tutor not make sure that their students have heard a correct statement before they leave? Just because the tutor has come up with some clever sentences, it does not follow that their students could then reiterate those sentences themselves. All they have done is assent to what they have heard. Tempting students to memorise sentences would be equally unproductive. In fact students leave lectures all the time with partial and incorrect impressions of ideas and techniques. What these tactics are offering is a means of exposing those misconceptions in a minimally threatening environment, where it is publicly acknowledged that the aim is to locate gaps and errors and to correct them. However, it is the students who need to be tempted into making assertions, not the tutor!

By being explicit about sense-making, students are more likely to engage publicly in forging meaning and in getting assistance to check conclusions they *have* made.

Tactic: Diverting Attention

Experts do not need to devote much attention to what they are doing when they are carrying out a familiar technique. They are able to watch for slips and to keep track of where they are going and what they intend to do with the result. On the other hand, students who are just learning a technique need to devote a good deal of attention to the mechanics of it. In order to move from novice to expert status, they need to reduce the amount of attention they need to devote to carrying out the technique. However, a collection of exercises designed to achieve mastery of a technique is more likely to attract attention than to divert it. Students are most likely to do each exercise in turn rather than try to see what is particular and what is general about each one. Consequently, it is helpful to have some techniques for reducing rather than increasing the amount of attention required.

Tactic: Ask students to explore something that will involve them in constructing and doing for themselves plenty of particular cases using a given technique. This has the advantage that the student is interested in the answers (because they feed into answering the principal question) and so attracts attention away from the details of doing the technique.

Tactic: Ask students to construct (and do) a question which shows that they know how to do questions of that type.

See also *Tactics: Particular – Peculiar – General*, p88; *Advising Students How to Make the Most of a Worked Example*, p97.

For example, instead of setting a number of integrals to do with a view to getting your students to choose which method to use, ask them to construct an integral that requires two (or more) specified integration techniques in order to find the answer. Instead of a set of exercises, all solving problems of a given type, change the task to that of exploring the possible types that can arise. Can students construct an example of an integration problem that might confuse other students into choosing an inappropriate method? Instead of asking whether some made-up functions are differentiable at certain points, ask students to construct functions that are (or are not) differentiable at specified points, perhaps with specified slopes and values. A minor example might be to specify an $f(x)$ in the interval $[0,3]$ so that it is differentiable everywhere, given that $f(x)=5$ for $x>3$, and $f(x)=1$ for $x<0$. Then perhaps add other conditions, such as having a maximum at $x=1$, and/or a local minimum at $x=2$. (See also *Issue: Developing Facility*, p180.)

Tactic: Exemplification

Tactic: Give one of your students a technical term, theorem title, or technique. Ask them to present examples of it to other students, who should then guess what the student was given to exemplify. It may be useful to have a second student as a checker, to make sure that the examples really are examples!

This is not nearly as easy as it sounds. Getting students to generate examples that are generic and illustrative takes time, as they need to be eased into the idea until it becomes a natural act for them to do by themselves. Having access to appropriate illustrative examples, or having the flexibility to modify a familiar example to make it do something, is perhaps the mark of a mathematician, but it is not an innate talent: it can be fostered and developed.

Time is always against the teacher. It may seem that the suggestions made here will take too much time away from other more pressing matters like covering content. However, once you have most students using one or more tactics for themselves, you will find that they will respond more fully and learn more efficiently. I am not suggesting that you use many different tactics in one session, either. Maximum impact can be achieved by the careful selection of one tactic when it is really needed, rather than using a whole sequence of tactics that merely divert everyone's attention away from the mathematics.

Collaboration Between Students

Employers seek staff that can work in a team, not brilliant individualists. Mathematical progress is sometimes made by individuals working alone, but usually they are part of a community or network of collaborators who are needed for checking proofs if not for inspiration and for possible ideas and connections. Thus learning to collaborate could be of benefit both to those intending to use mathematics to help them in another discipline, and to those intending a career in mathematical research.

In order to exploit the benefit of having others to whom to explain things, students can support each other by working together. They can do this not by copying out answers but by using each other as sounding boards and by explaining ideas to each other. It *is* cheating to copy what you do not understand, for you are cheating yourself; it is *not* cheating to learn from someone else's ideas. The crux is in what happens after the collaboration. If students focus merely on completing the task, then the task is not serving a very useful purpose; if students focus on using the task to learn, then its purpose is fulfilled.

Teachers have used a variety of tactics for trying to ensure that students learn from collaborating rather than simply hanging on to the coat-tails of other students (see Chapter 5 for specific tactics). In a tutorial, what is important is getting students talking to each other usefully, explaining what they think, offering examples and counter-examples, and explaining why something is or is not an example or counter-example.

General Tactics

Tactic: Being Mathematical

What do you do if a student asks you a question that you do not understand, or do not know how to answer? What if they ask you to do a question that you have not seen before?

Tactic:	Display your mathematical thinking.

Here are some examples.

- You can ask them (and also ask yourself) 'What Do I Know?' and 'What do I Want?' and write down their answers. These two simple questions rarely require much time to answer, but it is essential they be clarified and this often reveals a way forward when you are stuck. 'I Want ...' can often be developed, because one want may lead to, and hence be informed by, another more precise want, until one of them can eventually be dealt with.
- For each technical term you can ask for a definition that can be worked with, and perhaps an example similar to what is being asked for in the question. (These students can find in their notes or textbooks.)
- For each technical term involved you can ask for theorems which make use of that term, or techniques which involve it that they have met recently. (These students can find in their notes or textbooks.)
- You can ask for simple examples of mathematical objects that meet some or all of the requirements in the question.

This is what tutors do naturally when approached by a student with an unfamiliar problem. It is also what most mathematicians do when working on a research problem or when trying to understand a research paper. By displaying this behaviour publicly you orient students towards good mathematical practice. Eventually you are likely to see what to do, simply because you have more mathematical experience than they have. If not, you can always ask them to make partial conjectures (which is how mathematicians leave questions they have not yet resolved), and then tell them that you will find out for next time. Students can rarely be bluffed by pretending you know when you do not. More trust is lost by pretence than by honesty.

Tactic: Simplifying and Complexifying

Tactic:	Demonstrate how to attack a problem by simplifying it first, perhaps by lowering the dimension, changing the numbers, or examining a special case. Be explicit about how working on the particular enables you to see the general, to elucidate the structure, and so show you how to resolve the more complex problem.
Tactic:	Take a simple example and see how complicated you can make it: 'bury the bone' as deeply as possible, and then show the result to colleagues to see if they can recover it.

Tactic: Particular – Peculiar – General

Tactic:	Ask your students to construct or locate a simple *particular* example of some technical term, problem type, object satisfying the conditions of a theorem, or use of a technique. Then ask them to select an example that is *peculiar* in some way, perhaps because it exhibits some unusual features, or just because they think no one else will think of that particular example. Finally, ask them to construct as *general* an example as they can. (Adapted from Bills, 1996.)

- 'Give a particular, a peculiar, and a general example of a polynomial passing through the points $(1,0)$, $(2,1)$, and $(3,0)$.'

The interpolating quadratic is a particular example. Forcing the required values at the required points is a step towards the more general problem for any three given points and gives scope for different (peculiar) presentations of the same polynomial. Adding in a general polynomial multiplied by $(x-1)(x-2)(x-3)$ gives the most general polynomial solution. The method used can provide a stepping stone to the discovery of Lagrange's interpolation polynomial.

- 'Give a particular and peculiar example, and a general class of examples, of a function which is continuous everywhere but not differentiable at one point.'

If students participate in discovering that, starting from $|x|$, they can add on a differentiable function, translate by a differentiable function, and scale by a differentiable function, they may begin to appreciate the range of possibilities. They can then use these results to construct examples where the non-differentiability extends to several points, and then to infinitely many. Students often act as if there are only one or two pathological examples.

- 'Give a particular, a peculiar, and a general example of a sequence of positive real numbers which converges to 0 but for which the series does *not* converge.'

This draws attention to the factors that ensure the non-convergence of the series while retaining the convergence of the sequence.

As in the second and third examples, there is often no 'most general' example. The point is not only to appreciate this, but also to think about what factors can be altered, helping students to become aware of general classes of examples.

The move from a simple particular example to a peculiar example 'no-one else is likely to think of' often stimulates students to become aware of the range of possibilities from which they could choose. This is a precursor to becoming aware of, and then expressing, the generality from which they are selecting. Expressing generality seems trivial to experts (except when they are struggling to understand a new generality themselves), but it is far from trivial for novices. (See *Tactic: Expressing Generality*, p14; *Boundary Examples*, p14, 136.)

Tactic: Doing and Undoing

This is a major theme in mathematics (see p193), and it is a source of challenging problems and even of research questions.

Tactic: Instead of asking students merely to do repetitive exercises, insert one or more parameters in one of the examples and ask them to find the range of parameter values for which something happens.

For example, when solving inequalities of the form $x+2/x>c$, ask for what a the inequality $x+a/x>c$ has $x \geq b$. Similarly, ask your students to find all quadratic inequalities in x for which the answer is that x lies between a and b. This is a tactic which students could usefully be encouraged to adopt for themselves: generalising particular problems and finding the conditions under which the general problem is then still solvable.

Tactic: Take a simple computation, with its result, and ask your students to construct other similar questions with the same answer.

For example, 'doing' the integral $\int_0^2 (1-x)\,dx$ gives the answer 0; what features of the integral can be altered and still preserve the same answer (undoing)? Some students will not think of altering the limits, while others will not think of going beyond linear functions. When finding the most general linear example, it may surprise students that the arithmetic mean 'shows up', and trying to explain the coefficients for a quadratic function on the interval $[a,b]$ suggests connections with statistical moments.

Tactic: More strongly, after giving them a question to which they have to find an answer, give them some answers and ask them to construct the corresponding questions. Make sure you have explicitly done this several times with them in other contexts.

Here are some examples.

- Finding limits using L'Hôpital's rule can be seen as 'doing'; 'undoing' might involve asking your students to construct examples of functions which require a specified number of applications of the rule, or a class of 'functions' which have the same limit.
- Finding the intersections of a specific line and curve can be seen as 'doing'; finding all lines intersecting the parabola $y = 2x^2 + x - 1$ in two distinct points is a form of 'undoing' which stimulates students to think more generally as well as asking them for algebraic manipulation.
- Finding the limit of a sequence can be seen as 'doing'; finding the sets which can be the set of limit points of some set of real numbers is a challenging example of 'undoing', especially if you do not have a ready-made name for them.

Setting students a collection of straightforward 'doing' questions may seem as though it gets them to rehearse a technique and to develop facility with it. However, students often manage to 'do' questions without really thinking about them. They *work through* the questions individually but fail to *work on* them collectively. That is, they find an answer to each question but do not see the set of questions as being particular cases of a generality. By contrast, 'undoing' questions often require students to take the initiative to make up and carry out a range of examples for themselves, with the greater purpose of answering the 'undoing' questions. (See *Tactic: Student Generated Exercises*, p16.) It furthers the aim of getting students to examine many examples if students are familiar with the strategy of specialising; that is, making up specific examples for themselves in order to find out what is going on. (See *Tactic: Specialising, Generalising and Counter-examples*, p10.)

Tactic: Introducing Symbols

Tactic: When introducing a new symbol or technical term, repeat the 'long form' orally each time you use it, or point to where it is written on a board or poster. After a few times, start to refer to the long form more indirectly. When you feel your students are ready, pause after using the symbol or term and, when you have their attention, ask if anyone needs the long form used explicitly.

If you have been using the tactic *Pausing* (p50) then students will expect to do something when you pause, and by the act of pausing you will draw attention to the presence of something which needs to be worked on. Otherwise, technical terms may simply go flying by with students hoping it will all become clear later, as in a good detective novel.

The idea is to do for the students only what they cannot yet do for themselves. In other words, the aim is that your students take over the initiative from you. Here, gradually shortening the expanded form or meaning of a symbol or of a technical term gives students the cues they need to fill in the details for themselves. (See *Tactic: Directed – Prompted – Spontaneous*, p91).

Tactic: Directed – Prompted – Spontaneous

Tactic: Instead of always asking students the same sort of question ('Give me an example of …', or 'Tell me a theorem which uses this concept' for example), try giving them a less direct prompt such as, 'What question am I going to ask you?' or, 'What question do I usually ask you when you are stuck?'.

At first they may be unable to answer, but after a while they will begin to recognise that you have repeatedly asked them certain questions, and that they can learn to ask themselves the same questions. Eventually, you can focus on asking more deeply probing questions because your students will have internalised some of your questions into their own practice. The sequence of *directed* questioning, then increasingly indirect *prompting*, can eventually lead to students *spontaneously* using questions or prompts for themselves. You are teaching them to become more independent thinkers.

The same principle applies to using new technical terms (remind them what it means, but less and less fully over time), and to using a recently introduced technique (be explicit about fewer and fewer of the steps over time). The speed at which you withdraw support depends on your assessment of your students. (See also *Tactic: Introducing Symbols*, p90)

Tactic: Say What You See (yet again)

Tactic: One way to help students overcome uncertainty about notation is to get them to 'say what they see' in the notation, and then gradually assist them to piece together the whole expression from the details.

For example, in $\int_x^{2x} (1/t)\,dt$ many students may be so caught by the (implicit) presence of the ln function that they may literally not see the limits x and $2x$. If this is the case, they will make nothing of the result that the answer is independent of x. Asked to differentiate $\int_0^x (c+x)^2\,dx$, students may only see the x that they should not differentiate, or they may try to differentiate both. Sometimes students will persist in not mentioning one or more details, as if they are oblivious to their presence. This is entirely natural (people do it all the time), and it usually only takes someone else mentioning the feature to make them aware of it.

Underlying this tactic is the practice of 'saying what you see' to yourself when stuck on a problem, as a means of working out what you are missing. It is especially useful when applied to diagrams drawn by others. Diagrams that you draw yourself are usually clear to you not only because you have drawn them, but also because you are aware of the sequence in which the elements were drawn, and the reason for it. You know what is generally important and what is particular or peculiar to that drawing. You know what elements are related, and what the drawing is 'saying'.

Tactic: Drawing Diagrams

Tactic: When you find yourself with a picture in your head, consider inviting your students to imagine it too. That is, describe to them bit by bit what you are seeing, and invite them to imagine it for themselves (use of imperative language seems to work especially well, for example: 'Imagine a ...'; 'Add in a ...'; 'Bring in a ...'). Forbidding students to draw a picture while you are describing what you are seeing can help them to strengthen their powers of mental imagery. You can then get them to describe to each other what *they* are seeing, and only then to draw a picture.

More often than we realise, we use standard diagrams and standard examples which contain features that we know are irrelevant, but which students often subconsciously incorporate. Fischbein (1987) has called these *figural concepts* because very often the concept is bound up with properties deduced from figures. Classic examples of figural concepts include: all triangles have one edge parallel to the bottom of the page; all axes are horizontal and vertical, and always at right-angles; variables which are differentiated are at the end of the alphabet, and you only differentiate with respect to t or x, and occasionally y or z; all polygons are convex; A_5 is the only finite simple group; $x \to |x|$ is the only function not differentiable at a point.

Tactic: In situations where the previous tactic is unsuitable, you can simply draw a picture or diagram for your students, paying attention to the order in which you introduce elements, and making use of pauses between descriptions so students can keep up with you. The power of a diagram is not in the actual diagram itself, but in awareness of how it was constructed: the order of construction, the choices that were made, the freedoms and the constraints (what is general, what is particular, and what is necessary).

For example, if all lines are drawn with positive slope, or all quadratics are drawn crossing the x-axis with their axis parallel to the y-axis, then students are likely to assume that these are necessary features. See *Drawing Diagrams* (p42) for further advice on drawing diagrams.

Students have to make sense of diagrams just as they have to make sense of text. Diagrams have a great deal less grammar, and in a less familiar form. In order to experience how if feels for your students, interpret the following diagram for yourself as a statement about the equality of a sequence of ratios.

Task: Say What You See

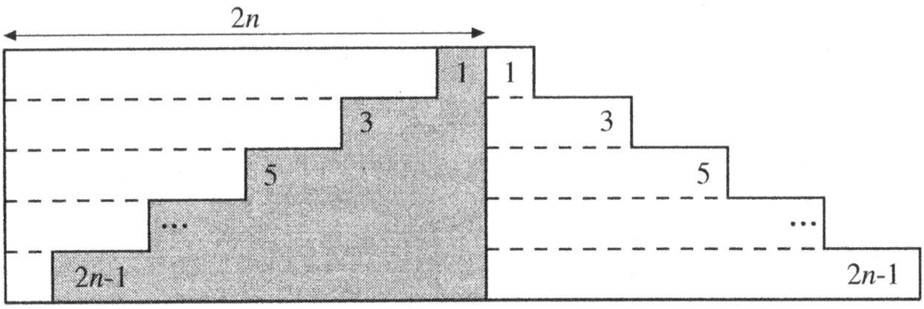

Try to identify what you do to make sense of this diagram.

This diagram was offered by Alfinio Flores (1998) as one of a series of 'proofs without words' used as fillers in the College Mathematics Journal. He also gave it a title, *Galileo's Ratios*, that may or may not be helpful, and provided a relevant sequence of ratios (see comment below).

Comment: Did you find yourself struggling at first? Did you try examining some special cases? If so, what did you specialise, and how? Did something trigger you to look at the overall shape, and perhaps to see three copies of the staircase? Were you fixated by seeing that it offered a visual proof that $\sum_{1}^{n}(2k-1)=n^2$?

Even if you have the ratio sequence in front of you

$$\frac{1}{3}=\frac{1+3}{5+7}=\frac{1+3+5}{7+9+11}=\ldots=\frac{1+3+\ldots+(2n-1)}{(2n+1)+(2n+3)+\ldots+(2n+2n-1)}$$

it is not immediately obvious how the diagram relates to the algebra.

You have to direct your attention to the components of the algebra (the numerator and denominator), see each of these as a sum, and then locate those sums in the diagram.

Once you have 'seen' the diagram through the algebra, or seen the algebra as an expression of the diagram, you will almost certainly want to generalise. The staircase is a special case of an arithmetic progression; can you arrange to get any ratio p/q in place of $1/3$? (See *Tactic: Doing and Undoing*, p89; and *Theme: Doing and Undoing*, p193.)

Task: Say What You See (again)

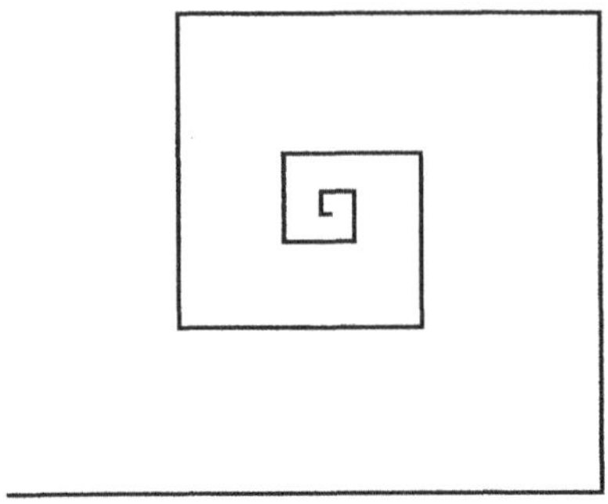

Try to identify what you do to make sense of this diagram.

Comment:

As an experienced mathematician, you will immediately recognise it as an example of convergence. You know that many different 'rules' for specifying the relative sizes of consecutive segments are possible, and that these will force the inward spiral to converge to some point. For example, if the nth line segment is θ/n times the previous length, then the point of convergence is $e^{i\theta}$ or $(\cos\theta, \sin\theta)$, relative to the starting point. However, convergence is so fast that it is hard to see more than the first three steps!

The purpose of the example was to give you experience of not immediately 'seeing' what the author is offering. This may help you to recognise what students may be experiencing when they do not immediately see what you are saying or drawing.

Tactic: Scrambled Proof

Tactic: Take an important proof, duplicate it, cut each copy up to isolate the different statements, then invite students to reconstruct the proof from the pieces. Get them to focus on how they work out where it starts, how each step follows from the previous one, and where it ends. Try breaking the proof in the middle of lines as well.

This is likely to help your students gain confident familiarity with features of the beginnings and ends of proofs, and how each statement follows from the previous one.

Tactic: Proof in Particular

Tactic: Give your students a proof applied to a particular case, then ask them to apply the same argument to another particular case, before stating and proving the general case.

For example, prove that $\sqrt{2}$ is irrational, and ask them to adapt it to $\sqrt{5}$, and then to generalise. When you catch yourself thinking, 'That will be too simple; it is obvious', then you have probably located a useful task for students, because what is obvious to you is not always obvious to them. If they too find it obvious, then you can press them to extend and generalise. (See also *Tactics: Getting Students Used to the Idea of Proof*, p53; *Presenting the Essence of A Theorem*, p52.)

Tactic: Catching 'it'

Students trying to express their ideas often use *it, this, that,* or *those* instead of specific nouns with definite articles. For example: '*You take it and differentiate it and then set it to zero and then …*'. Indefinite pronouns are often used to refer to something slightly or even completely different on each occurrence, and their use sometimes signals uncertainty if not confusion.

Tactic: Ask the student to pause, and to explain what the 'it' is before going on. After doing this a few times, many students will start to catch themselves using or about to use 'it', and often this can improve their articulation and hence their sense of the ideas.

On the other hand, pronouns are also powerful devices for use in summaries, and can even be seen as a precursor to symbolising. If there is some part of an expression you want to refer to, you can point it out and then refer to 'it'. Even more powerfully, you can label it with a letter. This chunking of bits of expressions, and using short forms to refer to more complex notions, is one of the ways in which mathematics uses symbols so powerfully.

A lecturer who uses indefinite pronouns frequently is likely to find that students have great difficulty in taking notes, because their note-taking cannot keep up with what is being said, so they do not know what is being pointed to physically when 'this term' or 'that element' is being used. Try to catch yourself using pronouns and then ask yourself whether your students are in a position to appreciate what is being referred to.

Tactic: Finding Out How Other People Think

Tactic: Try suggesting that students in a service course attend one session of a course for mathematics students dealing with the same topic, so that they can experience what it is like to enquire and probe more deeply into mathematics. (Suggested by Robert Baldino, who reported that this might have stimulated some students to begin asking 'why' questions in his class.)

Advising Students How to Study

This section offers some suggestions for advising students on how to study. Rather than simply describing study techniques to students and expecting them to try them for themselves, it is much more effective to engage students in a technique, and then to draw their attention explicitly to it as a potential study aid. In any case, most effective study techniques are derived from explicit use of characteristically mathematical ways of thinking and acting.

Techniques like these are most effective if students are taken through the actions more than once, and their attention is then brought to it as something they could do themselves (see *Tactic: Directed – Prompted – Spontaneous*, p91).

Halmos stressed many times the importance of examples in studying mathematics.

'If I had to describe my conclusion [as to the best method of studying] in one word, I'd say examples. *They are to me of paramount importance. Every time I learn a new concept I look for examples ... and non-examples. ... The examples should include wherever possible the typical ones and the extreme degenerate ones.*

A good stock of examples, as large as possible, is indispensable for a thorough understanding of any concept, and when I want to learn something new, I make it my first job to build one.

Counter-examples are examples too, of course, but they have a bad reputation: they accentuate the negative, they deny not affirm. ... [The] difference ... is more a matter of emotion.

Affirmative or negative, it's examples, examples, examples that, for me, all mathematics is based on, and I always look for them. I look for them first, when I begin to study, I keep looking for them, and I cherish them all.'

(Halmos, 1985, p62–64)

Most students will not establish the habits that Halmos formed for himself unless they are in the presence of people who exhibit those habits, unless they are taught by teachers who both use examples effectively *and* involve their students in example construction.

Tactic: Advising Students How to Study a Mathematics Text

A careful lecturer will have provided examples that meet a given definition, together with non-examples that highlight different aspects of the definition by showing them failing. However, examples provided by a lecturer are part of the wealth of experiences that often pass students by; they are part of the wallpaper. It is important and necessary to get your students to be active, so it is useful to get them to modify and create their own examples, for it is in the construction that students learn the significance of the details of a definition.

Suggest to students that when they read a definition, they should try to find or construct examples that fit the definition. Suggest that they practice testing whether appropriate objects do or do not satisfy the definition, in order to appreciate the various conditions stated. The purpose of finding an example is to become so familiar with it that it becomes intertwined with the definition; either can be re-constructed from the other if, under pressure of examination, you forget some details. (See Michener, 1978, and *Tactics: Specialising, Generalising and Counter-examples*, p10; *Boundary Examples*, p14, 136.)

Suggest to students that when they read the proof of a theorem, they should try to follow it step by step as applied to a familiar example, in order to appreciate the significance of each step of the theorem, and to see why the proof works. Trying to see where each assumption of the theorem is needed in the proof, by seeing where the argument would fail if that assumption were not made, is important mathematical work. You could illustrate this by regularly doing it on proofs yourself (Courant, 1981, Feynman, 1995).

Suggest that your students follow the advice of Halmos, who advocated that

'I believe that changing the notation of everything I read, to make it harmonious with my own, saves me time in the long run. ... [As] I keep changing the notation to my own, I get a feeling of being creative, tiny but non-zero – even before I understand what's going on, and long before I can generalize it, improve it, or apply it, I am already active, I am doing something.'

(Halmos, 1985, p70)

Tactic: Advising Students How to Make the Most of a Worked Example

Suggest to students that they connect each step of a technique with a theorem or principle with which it can be used.

Suggest that they could also try to use the particular to inform their sense of the generality expressed by the theorem or technique. One way to do this is to follow through the calculations of an example but not to perform any arithmetic. The idea is to track the calculations involving a particular number (using e, π, or $\sqrt{3}$ can be helpful) to see what happens to them. It is then usually a simple step to convert them into a generalisation, or, if this is not possible, at least to see the general through the particular.

Where an object is presented as an example of a structure (for example (**Z**, +, x) as example of an integral domain, $f(x)=\begin{cases} x(x-1) & \text{for } x \geq 0 \\ -x(x+1) & \text{for } x \leq 0 \end{cases}$ as an example of a function) suggest that your students work out what features of the example are necessary to make it an example, and what features could be changed. (See *Tactic: Student Generated Exercises*, p16.) Get them to try changing some of those features, to get a sense of the range of possibilities. (See *Tactic: Particular – Peculiar – General*, p88; see *Theme: Mathematical Powers*, p184.)

Tactic: Advising Students How to Make the Most of an Exercise

When doing a problem or exercise, the point is not just to answer that question, but to appreciate the range of questions of which it is typical. Recommend to students that they make up a similar question of their own (in order to appreciate what makes a question similar). Rather than doing a lot of similar exercises, advise students to do enough so that they 'know how to' or 'can write down a way to' do that kind of question, and can describe what constitutes 'that kind of question'. This is particularly helpful in helping students to become independent learners (see *Issue: Dependent and Independent Learners*, p177).

Students can be asked to create their own questions: an easy one, a hard one, and a general question of that type (see *Tactics: Student Generated Exercises*, p16; *Boundary Examples*, p14, 136). Alternatively, they can be asked for a particular, a peculiar, and a general question of that type (see *Tactic: Particular – Peculiar – General*, p88).

Tactic: Advising Students How to Integrate a Topic by Weaving a Story

Understanding is evidenced by having a 'story' to account for why things work and how they are interconnected. By a story, I mean an account or collection of ideas that provide justification, or a sequence of steps to be carried through.

Students could be encouraged to use various tactics such as *Linking Terms* (p84), *Story Telling* (p82) and *Arranging and Sorting* (p83) for themselves.

Tactic: Advising Students How to Learn How to Learn (Learning Files)

Writing is often a good way to learn, as Sterrett (undated) argues, even in mathematics. We are often called upon to express ourselves in writing as well as in symbols, and being able to communicate one's thinking is a valuable asset. Furthermore, some students need to learn that there is more to mathematics than doing calculations correctly. They too can be called upon to explain what they are trying to do in their calculations. Sterrett reports that refusing to accept submissions with inadequate textual explanation can be a way of forcing students to think about what they are doing instead of just doing it. However, he found that he had to stand very firm in the face of initial opposition from students unused to having to write in mathematics. He also got two students to referee each student's first draft of any extended work, exploration or investigation, because they learnt a great deal from commenting.

Other tutors ask their students to write a mathematical autobiography as a way of getting them to see how much their views of mathematics may have changed, perhaps at the beginning and end of their degree. Some tutors have tried asking students to write a short précis of the content of the sessions from the previous week as a way of getting them to make connections and to get an overview of topics. Students can also be asked to compare and contrast different proofs of the same theorem, or different types of proofs, indicating in what sorts of situations the different types might be most appropriate.

Many tutors get their students to keep journals. Some students' journals contain comments on their struggles and successes with problems (for example using a two column format), while some focus exclusively on the use of various structures and heuristics, as in Polya (1962), Mason *et al.* (1982) and Schoenfeld (1985), and some combine both. Some tutors take in this writing every so often and use it to engage in dialogue with individual students. Others find that the action of writing brief notes clarifies issues and helps students articulate their problems to tutors. If marks are assigned at first for including these sorts of comments, then students are more likely to do it seriously enough to see if it helps them, and hence to continue using them when they get stuck, long after you stop rewarding them with marks. (This is taken up in Chapter 5.)

At the Open University we encourage beginning students to keep a learning file, and, at the time of writing, there is growing pressure for all students in higher education to keep some sort of a record of achievement. Some students (often the weaker ones) find it very valuable, while others (often the stronger ones) find it does not suit them. Curiously, everyone struggles at some time, and that is when it is useful to track the struggle and to pay attention to how the difficulties were overcome. When students look back later, they will recognise both the improvements they have made, and the strategies they have developed for getting out of being stuck on a problem or on a concept. Polya (1957) called this *Looking Back*, and Mason *et al.* (1982) and Mason (1998) both offer structural advice on doing this. See also Meier and Rishel (1998).

Writing can be seen as a process of reflection, of inducing the writer to step back sufficiently from their immersion in a topic or a problem to become aware of what they find relatively easy and what they find difficult, and how they might get themselves out of difficulty in the future. Of course, it can only be effective if students are committed to it, which is much more likely if the tutor is also committed.

Tactic: Advising Students How to Recognise a Problem Type

Detecting the type of a problem is half the battle in solving it. *Arranging and Sorting* (p83) and *Linking Terms* (p84) are perhaps the best ways of coming to recognise problem types, and learning to look beneath their surface features to technique(s) which could be used to resolve them.

Since recognising a problem's type is crucial in examinations, this kind of activity can help prepare students for assessment. If you find students sorting inappropriately, you can choose to stress features that they seem to be overlooking.

Tactic: Advising Students How to Gain Mastery of a Technique

It is commonly believed that 'practice makes perfect', and that the best way for students to succeed is to do lots and lots of examples. However, many of the people who espouse this did not in fact do very many examples themselves. They only did enough to get a sense of how to do 'that kind of problem'.

Someone who has automated a procedure, who has developed facility with a technique, does not need to pay it very much attention when carrying it out. This allows them to keep the overall goal in mind and to watch for slips or surprises. In order to achieve this state of minimal attention, students need to have their attention drawn *away* from the technique they are trying to master (see also *Tactic: Diverting Attention*, p85). To this end, students can be encouraged to ask themselves what is the same and what different about a number of exercises, and invited to do as many as they feel they need to so that they can write down how to do a question 'of that type', or at least what constitutes being a question of that type. (See also *Tactic: Particular – Peculiar – General*, p88.)

Another approach is to offer challenging tasks that probe the boundaries of students' understanding. However, if a majority of the students cannot get started on a task, then there is little value and much harm possible for your students. (See also Appendix A.)

Structuring Tutorials

The origins of tutorials are rather obscure. Since Greek and Roman times, teachers have had sessions with two or three students, giving them plenty of opportunities to ask questions of their tutor. When there are no student questions to deal with, the tutor can probe their understanding, based on the homework they have submitted. This still happens in a few institutions, but for most the pressure of numbers means that tutorial classes are much larger. In 19th century America, students had to attend recitation sessions, in which they were expected to recite the main points of the previous day's lecture, and the name lives on in the USA today despite a change of mode. Most tutorials have at least 12 students, and often run to 30 or more with one or more graduate students in attendance.

For the most part, tutorials have now become exercise classes, in which students work through example problems and tutors assist them. The tutors circulate around the room advising, demonstrating, asking questions, and perhaps working through particular examples publicly on a board. Towards the end of the session, tutors may demonstrate solutions to the problems. However, tutorials can also be collaborative, and can demonstrate useful practices for studying and exploring mathematics. For example, when no-one knows an answer, several students can work with their tutor to locate it, with the tutor playing a guiding or supportive role, helping them to develop their ideas.

Seeing someone struggle publicly can be of real benefit to students who have the mistaken impression that mathematics springs from the pens of experts fully formed like a textbook. On the other hand, tutorials can turn into a confusing spectator sport if the tutor thinks rapidly about a question and then produces a neat solution. The key feature of a tutorial is that students should be active, with the tutor trying to enter their world, rather than the other way round. It is not even necessary that all tutorials be tutor driven or even tutor directed.

This section offers some suggestions for setting up tutorial groups and for structuring what happens.

Setting Up Tutorial Groups

Assigning students to tutorial groups at random can sometimes result in groups that are not very compatible in terms of their needs and desires. An alternative is to ask students to assign themselves to groups according to the grade they are hoping to get, or how much support they think they will require. This means that those who expect to need additional help can be put together and get that help, while those who are looking to be challenged can also be grouped together. Students seem to make most progress when they feel that their performance is among the best in their group. This also suggests changing the groups round every so often so that everyone has this experience!

As with most research in education, it is impossible to be definitive and precise about all the contextual factors within which one particular tactic will be effective. As is inevitable when dealing with human beings (who go up and down from day to day), it all comes down in the end to being sensitive to your students' needs, to your own mathematical awareness, and to the aims and requirements of the course.

Tactic: Self-selection

Tactic: Invite your students to assign themselves to groups on the basis of criteria such as:

- aiming to achieve a first, second, or third;
- wishing to do extra work to understand, comprehend, and appreciate the subject, to do enough work to have a sense of the important ideas and to master techniques, or to do just enough to pass;
- having considerable self-confidence, some self-confidence, or little self-confidence.

Finding criteria that students recognise but which do not simply reinforce their own impressions of their abilities is not easy. Sometimes a group of students will emphasise lack of confidence, and sometimes a group will be overly confident. The best policy is to describe what you expect from students who wish to work in different ways or with different intensities.

It is important that you respect your students' decisions and work to support what they want and what you perceive that they need. If they have strong preferences for one style of working while you think another would be valuable, then start by explicitly doing both (at different times). Over a period of time both their and your ideas about the best ways of working may change. Imposing an unfamiliar style or making a sudden change in how you interact with them is very likely to be greeted with suspicion and resistance. (See *Issues: Implications of Student and Tutor Expectations*, p166; and *Doing is Not the Same as Construing*, p168.)

Student Activities

Students can be asked to work individually or in small groups before they present their solutions to the class, or their tutor demonstrates how they would tackle the problem.

Students can present their solutions, either to homework or to problems worked on in the class, at the board. Alternatively, several students can write their solutions on different boards at the same time, for discussion and critique by their colleagues. If this method is used, however, it is very important that the critique is sensitive, and that it does not all come from the tutor.

Students can be assigned projects to work on individually or in groups, which involve them finding out something (such as the proof of a theorem, or discovering examples or counter-examples) for presentation to the class.

Some tutors require that students hand in work that they are going to present to the class in advance so that they have a chance to see and comment on it first. Others try to prompt students to critique the presentations themselves.

Peer Tutoring

In this approach, a group of 4 to 6 students is assigned a particular topic (or, in the case of an applied course, a particular model or case study) to read and understand. They then present their findings to everyone else, and field questions until everyone understands. Houston and Lazenblatt (1996a, 1996b, 1999) report that most of their students benefited from this structure, and also recognised the benefit.

The advantages of peer tutoring as part of a course are that students learn not to be dependent on a single authority (the lecturer), and to communicate their ideas to their peers. In the process, they often find that they have to adapt and change their own view of the topic, and sometimes even of learning and of mathematics. A possible disadvantage is that, while some students will find working in a group quite challenging and stimulating, others may find it difficult.

Another form of peer tutoring is to have students in one year providing back-up help for students in the years following them. When someone comes for help, the older students find themselves reconstructing what they know, and asking the younger students pointed questions, to which they need answers in order to make sense of the problem. This helps them to consolidate their knowledge and to develop their problem solving skills.

Adjusting the amount of assistance and support students are given can turn peer tutoring from a supportive environment in which students learn to reconstruct and re-present material to colleagues, into a pressured hothouse, as in the following example.

R. L. Moore Method

An extreme form of tutorial is the Moore method (MAA videotape 1966), developed by R. L. Moore for teaching graduate students in topology. In this method, students were forbidden to read any textbook or to seek any outside help. They were provided with a list of definitions and statements of theorems (sometimes not even definitions, which they then had to arrive at for themselves), and they then had to prove the theorems on their own. Tutorials consisted of students presenting their proofs, which were then critiqued by their colleagues and by the tutor. One of his principles was that if the tutor knew a shorter proof than the one presented by a student, the tutor kept quiet! Students who had not yet achieved what they thought was a proof of a theorem were expected not to attend the class when another student was presenting a proposed proof. One of the assumptions of his method was that there should be strong competition between students to be the first to prove a theorem, or to get the best proof, so collaborative work was banned. A number of famous and productive mathematicians went through this treatment, but it is clearly aimed at generating pure mathematicians. It could only be appropriate for students with a deep commitment to the subject, not for students in a compulsory course, or in service courses.

Students who could cope with it thrived on this approach, and many became famous mathematicians, but weaker students dropped out. However, softened versions of this approach may be more widely useful. For example, it may be helpful to base tutorials (if not the whole course) around students presenting solutions to set problems and re-constructions in their own words of proofs found in other sources. This can serve to make your students active with respect to working on mathematics while not requiring everyone to succeed at proving every theorem unaided.

Discussion of problem-based teaching and learning across different subjects can be found in Schmidt (1994).

Reflection

Task: Reflection

It is well worth pausing for a moment and making a brief note of any tactics that come readily to mind. What is it about them that sticks in your mind? Whatever it is that caught your attention may serve as a trigger to remind you to use that tactic at some time in the future.

Look back through the tactics in this section and select two that you might be able to use soon, and prepare yourself to use them by writing down what you think you would say and do, and then imagining yourself in a situation using them.

4 Constructing Tasks

'Challenge is the best teaching tool there is,
for arithmetic as well as for functional analysis,
for high-school algebra as well as for graduate-school topology.'

(Halmos, 1985, p271)

Introduction

In a sense, all teaching comes down to constructing tasks for students, because most students believe (however implicitly) that their job as a student is to complete the tasks they are set, including attending sessions and sitting examinations. This puts a considerable burden on the lecturer to construct tasks from which students actually learn. This chapter considers various reasons for setting students tasks, and considers the tension between tasks-for-teaching, tasks-for-diagnosis, and tasks-for-assessing, and whether there can be such a thing as an ideal set of tasks. As with the previous chapter, many of the tactics proposed here are analogues or variants of tactics that are mentioned elsewhere. However, a little repetition may be useful, since it provides an opportunity to revisit possibilities in a slightly different form and with a slightly different orientation.

Purposes, Aims, and Intentions

We set tasks for our students at the beginning of a course in order to diagnose their strengths and weaknesses, and any task can be used for this purpose to some extent. However, diagnostic instruments need to be constructed carefully if they are to reveal rather than obscure, and be helpful to students and informative for tutors. This issue is discussed in the first subsection below.

Most tasks students are set are intended, presumably, to stimulate them to think about topics, definitions, theorems, proofs, and techniques. However, what exactly do we want to achieve by setting such tasks? We want students to take the initiative, to internalise what they have been shown, and to gain facility in techniques (including proving and applying). We also want them to gain confidence in their mathematical ability and competence, and to experience what it is like to use mathematics well and to act as a mathematician. Is this what students expect as well, or do they have different ideas?

Task: Learning Theories

Make a list of the principal things you think students need to do in order to learn mathematics effectively. Now write down what you think your students would say in response to the same instruction.

In your next session, give each student a card and ask them to make their own list and then to hand it in to you.

Comment: A pragmatic view is that students expect that by doing the tasks they will somehow learn. Their job is to complete the tasks, by whatever means. Clearly it is not sufficient to copy answers from somewhere and submit them, but there are fine lines between the other options: learning by getting an idea from a colleague and re-constructing it for yourself in your own submission; copying what someone else has done but in the process coming to understand it; finding a similar example in a book and changing the numbers; and copying for the sake of the completeness of the submission without understanding.

We might hope that some students will include in their response things like 'reading around the course', 'constructing examples for myself' and 'reconstructing arguments for myself', but you are probably more likely to get things like 're-writing or going over my notes'.

The pragmatic response described above indicates an implicit view that is very widespread, even if rarely uttered publicly, and which may or may not be reflected in the pedagogy or beliefs of the teacher. It is the basis for an endemic tension between teacher and student desires (described more fully in *Issue: Implications of Student and Teacher Expectations*, p166). Certainly, the tasks you set will be taken by your students as exemplars of the kinds of tasks they will have to undertake in examinations. Thus one role for tasks is to provide students with the opportunity to practise a particular type of problem, and to become thoroughly familiar with it. However, if recognising the type is of importance, why not construct tasks that explicitly call upon students to do this? (See *Tactic: Student Constructed Exercises*, p16).

Modes of Interaction and Appropriate Tasks

Before deciding what sorts of tasks might be appropriate for different modes of interaction, it is necessary to consider the intended purpose of the tasks.

Task: Purposes for Tasks

Before reading on, make a mental note of different purposes you might have in setting tasks for your students. Then make a note of the different types of tasks that you use.

Comment: The purpose of setting tasks for students is presumably to enable them to learn something. However, there are many different ways of learning, as the previous chapters have suggested, and many different ways of encountering mathematical ideas. Among the many different purposes, you might have distinguished between tasks intended to achieve the following: revision of prerequisite ideas; introduction to what is problematic in a new topic; immersion in a topic for a period of time; posing the challenge of applying the ideas in a non-routine situation; and assessment of understanding. Of course, the same task can sometimes achieve several different purposes, but they may need to be crafted in order to be most effective.

The most common tasks that people think of are homework assignments intended to challenge students to get to grips with ideas and to master techniques, but there are also less obvious tasks such as sitting in a lecture and taking notes, and participating in a tutorial. Not all students are particularly clear about how to do these things! A more holistic approach acknowledges that students can take notes passively or actively, and that, since they need to be kept active, they are set homework and tutorial tasks, and even brief tasks to carry out during lectures. The following chart offers a starting point for delineating various types of tasks.

Mode of Interaction	*Typical Tasks*
Exposition	Students listen and take notes. Students discuss the topic briefly in pairs. Students do short, routine exercises in class and perhaps hand them in, but not for marks. Students read and make sense of a text or website, and use library resources (audiotape, videotape). Students rephrase what has been said, in their own words.
Explanation (tutorials)	Students ask (and learn to ask) focused questions about points that are unclear. You provide worked examples. Students display their own work to each other and to their tutor for comment and modification. Students construct illustrative examples, or boundary examples.
Expression (of students' own understanding)	Students make up problems for themselves which typify use of a technique, for example. Students construct examples to illustrate certain properties or conditions of a theorem, for example. Students express definitions and theorems in their own words, and articulate connections between concepts.
Exercise (practice)	Students do and re-do lots of exercises, and make up their own questions.
Exploration	Students do a starter task for homework prior to a lecture, exposing them to a typical problem that the topic resolves.
Examination (assessment)	Students do simple routine problems to begin with so that they gain confidence. Students asked to state definitions prove theorems, and provide examples to illustrate or contradict some statements. Students set routine problems similar to worked examples and homework. Students set a take-home or open-book exam with hard problems.

If students' activity is largely or even significantly driven by assessment, then it makes sense to use task structures which stimulate students to stand back from the doing of a technique and to get a sense of the structure and domain of applicability of the technique (a form of invariance amid change). They can also be encouraged to stand back from the proof of a theorem, and to get a sense of the ways of thinking used in the proof. If you want your students to do anything other than submit assignments, then the tasks you set ought at the very least to encourage (if not demand) that other behaviour. Examples of different kinds of tasks for different purposes are given in the next section.

Different Tasks for Different Purposes

Different sorts of tasks can be used at different times for different purposes. Each type of task can be used for diagnosis, for learning, and for assessment, though what is being assessed can be very different for different types.

Diagnostic Tasks

'This past decade has seen a serious decline in students' basic mathematical skills.'

(Hawkes and Savage, 2000, p1)

Diagnostic tasks are intended to inform the tutor of student strengths and weaknesses so that their teaching can be targeted accordingly. They are formative (for the students' benefit, not counting towards the final grade) as distinct from summative (counting towards their grade). As a result of the test, students can be informed of where they need to do further preparation, and they may be given special support in the form of texts to work through, problems to tackle, or tutorials in a learning support centre.

Consider, however, how you would feel if, on arrival at a new institution, you were immediately given a test and then told that you were inadequate or deficient in some topics or techniques. What if, as a result, you were placed in a group with others who were similarly 'deficient'. The test might be based on what the new institution considers to be 'basic', but it might bear little relation to your recent past experience. An experience like this might of course inspire you to 'get up to speed', but it might also feed your natural insecurities, already heightened by moving to a new type of institution away from home.

Thus diagnosis is by no means value and emotion free, no matter how much we would like it to be, and so it is wise to take students' feelings into account when designing diagnostic tests. For example, revealing strengths is a more positive approach than probing for weaknesses. Although this may sound like mere semantics, the way in which the results of the test are communicated to students can be crucial in how they respond to the diagnosis and to the plan of action that it entails.

Testing Effectively

The key element in diagnostic testing is to make sure that the test items do actually reveal strengths and do not just skate over the surface. When designing a test question it is easy to overlook the ways in which students might get the right answer for the wrong reason. For example, students who are asked whether $f(x)=x^5$ has an inflection at $x=0$ are likely to get the correct answer (yes) through incomplete reasoning ($f'(0)=0$). Although this may seem trivial, it is surprising how often students can get the right answer for the wrong reason, and, since you are unlikely to have time to look at all their working, you may be misled. By contrast, finding the relative extrema of $f(x)=x^5-7x^4+12x^3$ is likely to lead many students to assert there is a relative extremum at $x=0$ when there is not, if they have not checked the second derivative (and hence values near the critical point) as well. Thus they can get a wrong answer for an almost right reason. See also Avital and Barbeau (1991).

If the purpose of diagnostic testing is to inform as well as to test, then asking students what is wrong with a calculation may serve to alert them to errors. For example:

- A student claimed, incorrectly, that $f(x)=x^5-7x^4+12x^3$ has a local extremum at 0 because $f'(0)=0$. What was wrong with the reasoning?
- A student computed $\int_0^1(3x-1)(2x-1)\,dx$ by calculating $(3x-1)\int(2x-1)\,dx+(2x-1)\int(3x-1)\,dx$ as $(3x-1)(x^2-x)+(2x-1)(3x^2/2-x)$ and then evaluating this between 0 and 1 to get $1/2$. Why is this incorrect?

Here the wording more or less forces students to do the integral themselves, hopefully correctly, and to spot the error made (integration of a product is not like differentiation of a product). Those tempted to make the error are alerted to the fact that it is an error.

With all diagnostic testing, indeed perhaps all assessment, a student's reasoning is at least as important as their actual answer. Indeed, perhaps the most important thing you can offer your students is support in developing the habit of looking for ways to check their answers, not just at the end but every so often during a computation as well.

Over a period of time you can make a collection of student errors and misconceptions as suggested in Chapter 1 (see particularly *Tactic: Using Common Errors*, p10), and use these as test items. Problems on which other students have foundered previously when doing assignments, exams, and other tasks, are an excellent source of material.

There is a slight caveat, however. Simply extracting the parts containing the errors from longer computations to make a diagnostic test may not be as effective as you expect, since students will know that it is a diagnostic test. They will be on the lookout for traps, and be more careful than usual. They will be far more likely to slip up when they are immersed in a longer computation and only half attending to their algebraic manipulations. It is much easier to make a slip on something 'routine' when your mind is on some more complex task than it is when your attention is fully on the routine task. Thus students who can get most of a diagnostic quiz correct might still make those errors in the midst of a longer problem. It is when attention is only partial that many misunderstandings emerge. This is the negative side of the tactic *Diverting Attention* (p85), since flawed calculations can be automated almost as easily as correct ones. Here are some further examples, contrasting an extracted error with what may happen in a context in which it might arise.

- Asking students to expand $(a+b)^2$ or to factor a^2-b^2 may trigger correct responses where asking them to expand $(3a^2-2b^3)^2$, $(\sin x+2\cos x)^2$ or factor $32x^4a^6-18y^8b^4$ may not.
- Asking students to use a probability density function to read off some probabilities may reveal uncertainties between distributions and density functions, but these uncertainties are more likely to emerge in the midst of a more complex task in which the probabilities are needed for some other reason.
- Asking students to integrate a number of functions may be less informative than asking them to choose options that describe what one looks for in choosing a method to use.

What exactly is being tested in the first example? If you want your students to recognise complex presentations of the difference of two squares, then perhaps it would be more effective to get *them* to try to 'bury' the difference of two squares in different ways (see *Tactic: Student Generated Exercises*, p16). If your aim is to ensure your students always expand a bracket correctly, it might be more effective to get them to develop the habit of checking calculations, for example, using a tactic such as *Specialising, Generalising and Counter-examples* (p10). Using this tactic, you explicitly ask them to specialise or to construct a counter-example, and then, over time, prompt them more and more indirectly until they have taken on the process of checking for themselves (see *Issue: Dependent and Independent Learners*, p177).

If you want to learn what it is that leads students to make errors, rather than just the fact of the errors themselves, then a diagnostic test may not provide very much information. Indeed, until you know what leads students to write what they do you are not in a very good position to construct effective diagnostic tests. To find out more, you can appeal to a learning support centre (where there is one), to the literature (which is extensive), and to the students themselves. For example, having detected some classic errors you could interview students about them. However, since interviewing is not entirely straightforward, and students may be over-awed if their lecturer starts asking questions, it may be better to use other methods to gather information and at the same time to influence students, as in the next tactic.

Tactic: Using Students to Learn About Students

Tactic:	Try getting third-year students, particularly those interested in the possibility of becoming teachers (at whatever level), to interview first-year students about classic errors and confusions as part of an assignment.

Your undergraduates are more likely to be able to empathise with first-year students, and be able to supplement what they hear with their own experience. It will also be useful for them to experiment in trying to extract information from other students about teaching and learning.

Another approach that can be used in every course is to get your students to construct questions which they think will expose confusions or difficulties for next year's students (see *Tactic: Student Generated Exercises*, p16).

Tactic: Using Students' Reflections

Tactic: Ask your students to keep a reflective diary in which they comment on what they think each session was about, and describe the difficulties they are having with the content and with the homework assignments. You can arrange to look at these overnight every so often, in order to augment your own sense of what they are finding difficult. The diaries may also be useful to them later, when they look back to see how much they have learned. In a sense, such diaries can be seen as the mathematical equivalent of a lab book in physics or engineering.

For some students, the notion of a diary may be less helpful than that of a notebook in which they keep a record of significant ideas they have just met, or important ways of thinking that they have experienced or run up against. The notebook then builds up as a record of their expertise.

Testing Efficiently

The second key point about diagnostic testing is that reading through hundreds of diagnostic test responses is not an inspiring way to begin your term. Ideally, you want a test that can be administered easily and that tells you and your students what you and they want to know (Hawkes and Savage, 2000). Fortunately there are several such systems available.

- CUE (see *Websites*) consists of a diagnostic multiple-choice test for service mathematics students, a weekly monitoring test for calculus students, and a grading test. (Jackson, Fiddes, and Beevers, 2000).
- TAL (see *Websites*) provides multiple choice computer-based tests for first year engineering students, which can be tailored by including or excluding material as appropriate so as to be suitable as to question type, difficulty, and time taken to complete. (Barry, Williams, and Wilson, 2000).

What remedial action might you prescribe for your students? They will benefit most if they are informed of the topics on which they need specific help, but it might be unwise to put too much emphasis on individual students. In Chapter 1 it was suggested that the tactics *Using Common Errors* (p10) and *Specialising, Generalising and Counter-examples* (p10) could be used briefly at the beginning of a lecture or tutorial as a warm-up activity. Some lecturers have found this to be very effective in countering classic confusions.

Not all testing has to be for marks, or result in students being reminded of their weaknesses rather than their strengths. For example, formative assessment gives students a chance to submit an outline or some rough working, for comments and suggestions from their tutor, before they hand in the final assignment. This is most often used when the assignment is a project or an essay. Students can offer their current state of thinking and receive advice on what to do to improve it. Just as with diagnostic feedback, the form and tone of your response can make a big difference to whether the student completes the assignment with interest and enthusiasm, or tries to get away with the bare minimum required.

Tactic: Commenting on Student Scripts

Tactic: Find something to praise, if only the fact that some conjectures have been reached and perhaps even rejected or modified. Offer suggestions rather than hints (see table, p32). Remember that, however inadequate the submission in your terms, it probably represents considerable time and energy on the part of your student. (See also Chapter 5.)

It is important that diagnostic assessment is not seen as simply probing for errors and confusions, as once students are in the mood of trying to hide what they do not know, the whole exercise works against itself. If, on the other hand, students feel that they are going to be rewarded for what they can do, then they may actually perform better as a result.

Tactic: Promoting Self-checking

Probably the most useful thing students can learn is that they should check their own work. You can help them in at least the following two ways.

Tactic: Point out places in your students' scripts where they could have checked a conjectured answer. Pause when doing a long calculation and publicly check that the current state conforms with one or more known particular cases (and point out to your students that they can do this for themselves as well).

The result of this tactic will hopefully be to stimulate the awakening of the inner monitor that every successful user of mathematics has, which checks manipulation subconsciously, as well as, every so often, asking 'Why are we doing this calculation?'. Such an inner monitor can also put the brakes on a tendency to rush on to the next task when a calculation seems near completion and ask, instead, for some way to confirm or check the result before leaving the task.

Start-up Tasks

Start-up tasks are useful as an introduction to a topic, perhaps as a short homework exercise, to bring your students up against the sort of problem that the topic is going to show how to resolve. Here are some examples that can be used to prepare for the concept of limits.

- The sequence $\{1, 1/2, 1/3, 1/4, \ldots\}$ gets arbitrarily close to (has as its limit) 0. The sequence $\{1, 1/2, 1/3, 2/3, 1/4, 3/4, \ldots, 1/n, (n-1)/n, \ldots\}$ has two subsequences, one of which gets arbitrarily close to 0, and another that gets arbitrarily close to 1. Can you construct a sequence that has subsequences that get arbitrarily close to three, four, or n distinct values?
- A similar task could be the following. What points can subsequences of the set D get arbitrarily close to, if D is formed by placing a decimal point in front of all whole numbers, so $D = \{0.1, 0.2, 0.3, \ldots, 0.9, 0.10, 0.11, \ldots, 0.99, 0.100, 0.101, \ldots\}$? (Idea due to Bob Burn.)
- Another similar task is to explore the value of $(1+1/n)^2$ as n gets large.

The aim is to stimulate students to experience limiting processes outside the rarefied context of convergent sequences, in order both to expose the value in isolating convergent sequences as a suitable topic of study (typical mathematical thinking) as well as, in the case of the first two, to set up topological ideas such as density for the future. Here are some further examples from other areas of mathematics.

- (To prepare for characteristic parameters of distributions.) How far apart can the mean and the mode be for a specified (finite) number of data points on the interval $[0,1]$?

The aim is to get students to think of a (finite) distribution not as a fixed entity but as one of a variety of different possibilities with different characteristics, and to think about what sorts of information the mean and the mode provide.

- (To prepare for a discussion of modelling.) If you take a smooth stick (such as a metre rule) and support it at each end by one finger, then slowly move your hands together, you will find that the stick appears to move irregularly. Explain.

The aim is to give students the experience of modelling a simple situation for themselves, one that they can actually experience.

- (To prepare for derivatives.) Describe what values the slope of a chord can have if one end is fixed at particular point on a curve, while the other is free to wander along the same curve. (Try it on some specific curves.)

The aim is to get students thinking about the slope of a chord, and what it does as the wandering end gets closer and closer to the fixed end.

Mid-topic Tasks

In mid-topic you want to stimulate your students to explore around the techniques and ideas you have introduced, constructing examples to show that theorems are (or are not) well-defined, illustrating when a technique is applicable and when it is not, and generally encountering the wrinkles that may arise when a technique is used in practice. For example, they may discover points such as the following: using integrals to find areas requires care when the function is negative; finding the angle between two vectors fails when one of the vectors becomes the zero vector; taking limits of two variables simultaneously may depend on the route that is taken; the values of $0/0$ and 0^0 are context dependent and have no single consistent value. Here are some further examples.

- A set S has p as a *point of accumulation* if every open set containing p also contains points of S other than p (where p may or may not be in S). Some sets contain all of their accumulation points, while others do not. Construct some sets in the plane that contain all their accumulation points, and others that contain none. What is the same and what is different between the two cases? Specify some sets on the line and in the plane that have no points of accumulation.
- The set $\{3, 9, 15, 21\}$ forms a group under multiplication modulo 24. It is isomorphic to the group U_8 formed by the set $\{1, 3, 5, 7\}$ with multiplication modulo 8, by multiplying every element of U_8 by a constant (3). What other groups can you construct in an analogous manner?

The aim is to get students to construct their own examples, and, in validating their examples, practice the use of the definitions of accumulation point and isomorphism. In the second case, students will also become familiar with the important role played by the identity.

- One student decided that $\lim_{x \to \infty}\left(\sqrt{x^2+x}-x\right)$ is 0 on the grounds that both terms are unbounded. Another student de-rationalised the denominator by multiplying the expression by $\left(\sqrt{x^2+x}+x\right)/\left(\sqrt{x^2+x}+x\right)$, and decided that the limit was $1/2$. Another expanded the square root using a power series and also got $1/2$. Which, if any, was correct?

The aim is to draw students' attention to the effects of the definition of continuity.

- A string is hung over a horizontal cylindrical rod. One end supports m_1 grams and the other end m_2 grams when the string is on the point of slipping. Find the coefficient of friction between the rod and the string. (Based on Wright, 1930, p70)

The aim is to alter the standard situation of a mass on a flat surface, to see if students appreciate the basic idea.

One of the temptations in constructing tasks is to provide too much information, thus reducing the challenge from a significant exploration to a sequence of subtasks that the student can carry out without realising what they have to do with the main task (Nardi, 2000b). For example, the group theory problem above is a particular case of a general theorem, that every group formed as a finite set of numbers under multiplication modulo arbitrary m can be displayed as a isomorphic image of a subgroup of the set of residues relatively prime to M, under multiplication modulo M, for arbitrary M. Once aware of the theorem, it is tempting to try to lead students towards the result. As an extension task it could perhaps be set simply as a theorem, but what sort of structural support might you offer to make it useful as a mid-topic task?

You could offer a particular example, as was done in the group problem above, or you could offer several examples. You could provide just the groups themselves, or the groups together with the isomorphic groups, and you could even specify the multiplicative isomorphism. You could also ask students to locate the identities in each of the examples, and to use them to construct (or verify) the isomorphism. To support students in proving the general theorem you could structure an argument in a particular case for them. However, once you embark on offering some help, it is terribly easy to be seduced into offering a great deal, removing the challenge altogether! You can end up with a sequence of much easier tasks that your students can carry out without appreciating the general problem or the process of breaking it down for themselves.

Extension Tasks

An extension task is designed to attract students who are confident about routine tasks. However, a revision task for one class could be an extension task for another, depending on what has been presented in lectures. For example, with limits, the notion of accumulation points could be used as extension if it is not part of the syllabus. An extension task on accumulation points could then be something such as the following.

- For the real line, which types of subsets can be the set of accumulation points for some subset of the line? Must a point of accumulation of the set of points of accumulation of the set S itself be a point of accumulation of S?

Note that the students are being asked to characterise a collection of objects that arise from a definition. This is not only a typical mathematical move, but also a useful one to get students to undertake, to help them think this way themselves.

- Having decided that $\lim_{x\to\infty}\left(\sqrt{x^2+x}-x\right)$ is $1/2$, specify the most general function you can find $p(x)$ such that $\lim_{x\to\infty}\left(\sqrt{x^2+p(x)}-x\right)$ exists. Specify and justify the limit.

Note that the students are being asked to think about a technique (de-rationalising denominators involving a square root), to consider the class of functions to which it applies, and to construct for themselves a family of functions. They are also being asked implicitly to explain how the answer $1/2$ relates to the coefficients in the original function.

- A string is strung alternately over and under some identical parallel cylindrical rods and, when the ends are attached to masses m_1 and m_2, the string is just on the point of slipping. What is the coefficient of friction between the rods and the string? How else might the string and the two masses be supported by the rods and how would the calculated coefficient of friction change?

The aim here is to suggest a direction for generalising an earlier task.

- A table stands on a smooth flat floor. What is the distribution of the mass of the table among each of the legs, and why? Justify any additional assumptions you have to make. What other situations can you think of that might require additional assumptions?
- Newton's cradle consists of a sequence of identical ball bearings hung in a line from a support. One ball is swung out of position and then released. Confirm theoretically the observed behaviour, that the ball at the other end swings away and the others remain still (in a well constructed cradle). Explore what happens if you have three ball bearings of different sizes in the cradle, hung so that their points of contact are in a line.

The aim is to get students to recognise under-determined situations and the need to make further assumptions in order to be able to construct a mathematical model.

The purpose of these examples is to illustrate the kinds of tasks that you could set for your students in order to extend their thinking. In the process of setting tasks, whether for start-up, mid-topic or extension, it is often handy to choose examples that have integer answers. Finding classes of these can lead to problems that are mathematically interesting in their own right (see Cuoco, 2000, for a general discussion, and Nering, 1970, for a collection of integer matrices with integer inverses). However, be careful not to overdo it, otherwise your students may be worried if they get non-integer answers to examination questions!

Tactic: Characterising

Tactic: Instead of, or in addition to, asking students to do a collection of exercises, ask them to characterise the sorts of answers that they could get as answers to that type of question.

Examples of this tactic could include the following.

- Characterise the conditions in a problem like [specified] that enable a specific technique or theorem [specified] to be applied.
- Characterise the 2 by 2 matrices that can be the square of a matrix over the rationals (or over the reals). Characterise the 3 by 3 matrices that have [specified] eigenvalues.
- Given a continuous function f on $[0,1]$ to the reals, the set of points of $[0,1]$ on which f is non-negative is a closed set, in fact the union of closed intervals. What unions of closed intervals can be such a set?
- The integers that can be written as the sum of two squares are closed under multiplication. Characterise numbers that are the square of a number which is itself the sum of two squares in terms of Pythagorean triples.
- Characterise in terms of their edge lengths all quadrilaterals whose diagonals intersect at right angles.

Characterising questions require the student to specialise by constructing a range of objects for themselves having various specified properties, raising questions such whether there is an object with a particular property. These may have to get more and more complicated before the student 'sees' what sorts answers can arise (in the examples above, what matrices, what sorts of unions of closed intervals, what sorts of numbers). Thus the students take the initiative to construct objects, and in the process enrich their sense of the range of possibilities covered by such important concepts as 'continuous functions on $[0,1]$'.

Characterising problems are a manifestation of the mathematical theme of *Doing and Undoing* (p193). When you know how to 'do' a class of problems, you can ask the 'undoing' question of how to characterise the solutions that can arise from that class.

In applied mathematics, extension tasks could take the form of exploring the effects of modelling assumptions, as in the following example.

- From the pendulum equation $\ddot{\theta} = -\omega \sin\theta$ one can find that $\dot{\theta} = (2\omega)^{1/2}\left(\cos\theta(t) - \cos\theta_0\right)^{1/2}$. The equation can also be approximated to $\ddot{\theta} = -\omega\theta$, yielding $\dot{\theta} = \omega^{1/2}\left(\theta(t)^2 - \theta_0^{\,2}\right)^{1/2}$. For $\theta_0 = \pi/2$, show that the error in the maximum velocity of the first swing predicted by the approximation is of the order of 15%. For what values of θ_0 does the approximation to the maximum velocity stay within 1% of that given by the full equation?

Mathematical Literacy Tasks

Mathematical literacy is complex, because it involves making sense of text, decoding symbols, and interpreting diagrams. Students also need to learn to express themselves, to put things in their own words (and symbols, and diagrams). In this process, the tutor does not have to be the only audience: students can present to each other (and receive comments from their colleagues), construct posters for display in the department, and write web pages. What is important is that they have the sense of an audience to whom they are presenting their ideas. See Sterrett (undated) for a wide-ranging discussion of this issue with strategies for attracting students to 'write in order to learn'.

Working with Text

We want our students to learn to read and make sense of written mathematics, so it makes sense to get them to read a short paper or to explore part of a website, to make up their own illustrative examples and counter-examples of what they find, and to present the results and examples in their own words. They may even be able to extend them in some way, at least to the extent of making sensible conjectures if not obtaining proofs.

Some tutors spend part of their tutorial time getting their students to present reconstructions of theorems and proofs, or examples which illustrate particular concepts, to each other for criticism (in a mathematical atmosphere of course). This way, their students learn to look for gaps in arguments and to consider how well an example illustrates a concept, theorem, or application of a technique. An extreme form of this is the R. L. Moore method (see p102).

Getting students to write coherently and clearly is a non-trivial enterprise. Good writing stems from wanting to express oneself, not from other people's pressure to write. Students have to want to express themselves before they will take it seriously. That desire often arises, for a mathematician, from the sheer excitement of discovering (what appears to be) a result or a counter-example. One approach, therefore, is to offer some incorrect conjectures for them to explore, contradict, refine and modify. Ask your students to justify their alterations, and to present them to each other, or on the web. One of the features of the web is that, when students know that their work will be placed on it, they are conscious of having an audience (potentially anyone in the world!) and so put more effort into their presentation.

Here are some sample tasks (see also Chapter 6).

- Rosenbrock's function $f(x)=\sum_{k=1}^{n-1}\left(100(x_{k+1}-x_k)^2+(1-x_k)^2\right)$, also known as the banana function, was designed to test the speed of convergence of algorithms for minimising functions of several variables. Show that f has a global minimum of 0 at the point with all coordinates equal to 1. Estimate the change in the minimum value and the point at which it is attained for the case $n = 2$ if the coefficient of x_2 is changed by ε. By searching on the web, find out about the history of this and related functions, and the reason for the name *banana.*

Here the students are invited to consider the simplest case, and to find out about the function on the web where there are pictures and references to original articles.

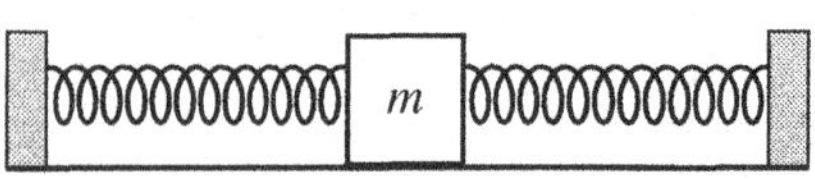

Two springs are attached to opposite sides of a mass m with their other ends fixed, as shown. Suppose the natural frequency of such a system is ω, and that doubling the stiffness of the first spring changes the natural frequency to ω_2, while doubling the stiffness of the second spring changes the natural frequency to ω_3. Find a condition on ω_1, ω_2, and ω_3 for which this is possible, and show that it can indeed be satisfied (Groetsch, 1999, p160). Generalise!

Here students are invited to reverse the usual form of the problem, which is yet another example of *Doing and Undoing* (see p193).

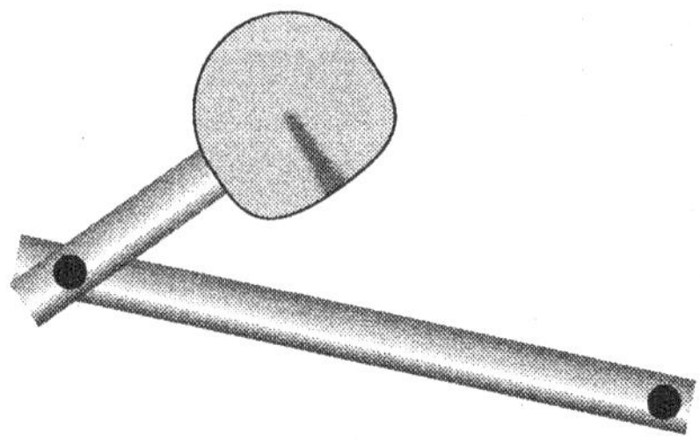

A fairground ride consists of a seat on one end of a short arm which rotates with angular velocity ω_1 about a pivot, which is on the end of a second arm that is itself rotating about its other end with angular velocity ω_2. Suppose a second ride could be made which is effectively the same as the first except that the arms' lengths and angular velocities are interchanged (so that the longer one still has angular velocity ω_2). What differences would riders experience between the two rides?

Here students are expected to model the situation and to discover that the two rides are identical as far as the rider is concerned because vector addition is commutative, illustrating the theme of *Invariance Amid Change* (see p192).

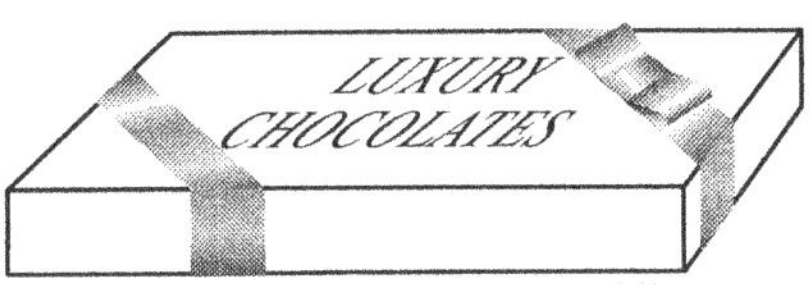

A box of chocolates has a ribbon tied in the usual way, going over one corner and under the next, over the third and under the last. For given dimensions of the box, what positions of the ribbon on the box minimise the length of the ribbon?

Here students are expected to have had experience of tying or releasing ribbons tied in this fashion, but not to have realised that the length of the string is independent of the path of the ribbon, another example of *Invariance Amid Change* (see p192).

- Compare the relative speeds of a gramophone stylus at the beginning and end of playing an LP (33 1/3 rpm, 15 cm outer radius, 6 cm inner radius), and the relative speeds of a CD which changes its speed so that the head is always reading at the same speed (1.25 m/s, maximum radius 6 cm and minimum radius 2 cm). Assume that the motions in each case can be approximated as circles rather than spirals.

Working with Symbols

Many of the tactics proposed in the earlier chapters are concerned with getting students to work with symbols. If they tend to be put off by strings of symbols, it often helps to show them how to break an expression down, to adjoin words and diagrams, and to try out particular cases. For example, the definition of a limit

$$\forall \varepsilon > 0, \exists N > 0 \text{ such that } \forall n > N, |u_n - L| < \varepsilon$$

can be deconstructed along the following lines.

Show from the definition that $\lim_{n\to\infty} \frac{1-n^2}{2n^2+n+2}$ *is* $-\frac{1}{2}$.		
$\forall \varepsilon > 0$	I can choose any ε that I like, but the intention is that it should be small, and that all that follows works no matter how small an ε I choose.	Try $\varepsilon = 1/10$.
$\exists N > 0$	There must exist (I have to find) some number N (probably large),	I want to find a suitable N.
such that $\forall n > N$	such that no matter what n is chosen, provided that it is larger than N,	
$\lvert u_n - L \rvert < \varepsilon$	$\left\lvert \frac{1-n^2}{2n^2+n+2} - \frac{-1}{2} \right\rvert < \varepsilon$.	I need to know for which n $\left\lvert \frac{1-n^2}{2n^2+n+2} - \frac{-1}{2} \right\rvert < \frac{1}{10}$, so I do some calculations to reveal that $n \geq 4$ will suffice. This means that $N = 4$ would do for this case.
	I then repeat the computations with a general ε to find that $n \geq 3 + 1/4\varepsilon$ will always suffice.	I have to verify that my choice of N works, at least in the general case, and for that I might use the particular to help guide my calculations in general.

Although all of this is obvious to an expert, it may help students who have a tendency to panic when something looks complicated, and who are tempted to avoid or ignore bits they do not understand or are uncertain what to do about.

Working with Diagrams

As mentioned in the tactic *Say What You See* (p20), there is a great deal that can be done with diagrams. However, not all diagrams are equally easy to use. Some of the principles that have been identified as aiding interpretation are the following (Tufte, 1983, 1990, 1997, Casselman, 2000).

- Provide sufficient structure without cluttering.

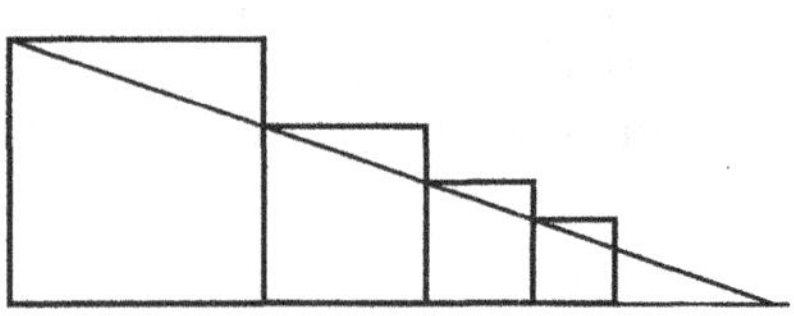

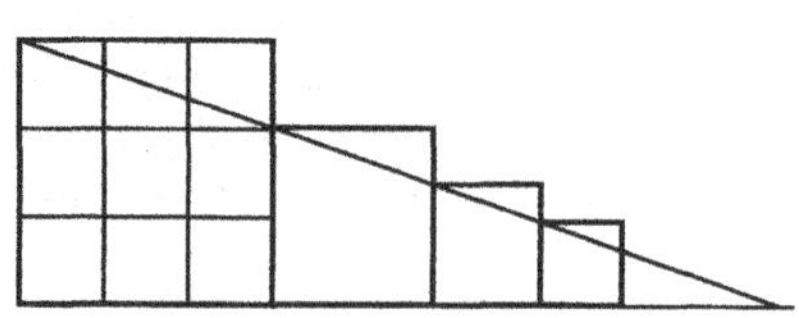

Here the two diagrams can each be read as a way to calculate the sum of an infinite geometric progression. The first suggests generality, but in so doing obscures some of the construction, while the second supports the construction in a particular case but is then open to generalization (Casselman, 2000). For this reason, it is perhaps better to use both diagrams together (particular, then general).

- Since there are usually a number of features in a diagram to which you want to attract attention, try using several diagrams, each emphasising a different aspect.
- Diagrams should tell some sort of a story. Sequences of diagrams are better for this (and can be presented as animations). Thus the diagram above tells a more obvious story if it is redrawn as the following sequence.

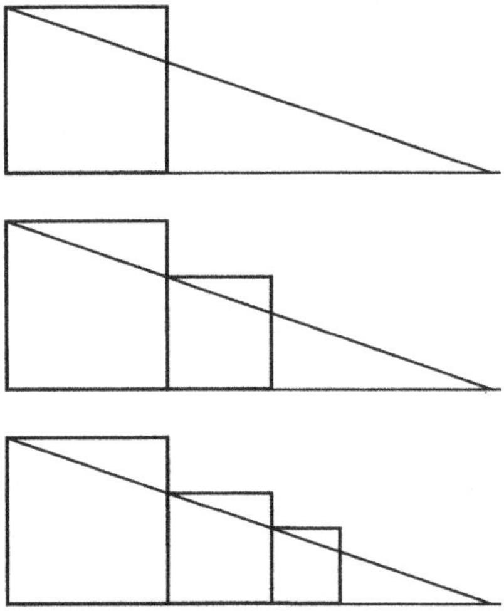

- It is rare for there to be too many diagrams (whether in exposition to students or in a research paper!).
- Try out your diagrams on other people; it is amazing how interpretations vary!

A large collection of 'proofs without words' can be found in the pages of the *Mathematics Magazine*, and in Nelson (1993, 2000). It can be very productive to work on using diagrams with students, getting them to comment on the effectiveness of the ones you use, and to construct their own variants which you could use the next time. In the process, they will hopefully learn to work on diagrams for themselves.

Revision Tasks

The purpose of revision is to draw back from the details and make sense of the overall picture, as well as to develop facility. Various suggestions have already been made for this purpose. Among them are the *Tactics: Making an Example be Exemplary*, p29; *Reconstructing and Re-expressing*, p25; *Story Telling*, p82; *Student Generated Exercises*, p16.

Assessment Tasks

If the assessment tasks will require students to use a technique or to apply a theorem from the course, then a standard approach to revision is to offer them more routine questions which can be done by recourse either to the text, lecture notes, or worked-examples used as templates in the lectures or tutorials. Students may gain something from finding the correct template and then changing the numbers, but it is more likely to be superficial pattern spotting than any deep connection with the underlying ideas. (See *Tactic: Student Generated Exercises*, p16.)

However, if the tasks are too hard, so that most students cannot make a start on even the first problem of a set, then they will very quickly become disillusioned and depressed. Although students who understand a topic can do hard problems, it does not follow that giving students hard problems will make them understand. There is a fine balance to be struck between sufficient and excessive challenge. (See *Issue: Challenge – Insufficient or Excessive*, p166.)

One way to balance this tension is to educate your students to modify, extend, vary and generalise the tasks they are set, as part of the development of their study skills. A student who takes a fairly routine exercise and then modifies or generalises it is more likely to be motivated to work on that new task. (See *Tactics: Specialising, Generalising and Counter-examples*, p10; *Particular – Peculiar – General*, p88; *Student Generated Exercises*, p16.)

Another way to achieve the right balance is to make use of the theme of *Doing and Undoing* (p193). Instead of setting one or more questions involving a technique, set a problem in which the answer is given, and ask your students to work out what the initial conditions were to which the technique was applied to get that answer. This is likely to involve them in constructing and resolving their own examples, rather than simply doing ones that you have set for them.

For example, instead of specifying a rational polynomial and asking for asymptotes and stationary points, ask them to find the values of a and b in $y = \dfrac{x^2 + 3x + b}{x + a}$ $(x \neq -a)$ for which $y = x + 4$ is an asymptote and $x = 2$ is the position of a stationary point. As a greater challenge, ask them for the most general rational polynomial of the form quadratic over linear that has the given asymptote and stationary value. An even greater challenge would be to ask them to write down an algebraic description of all such rational polynomials (see *Tactic: Particular – Peculiar – General*, p88).

Similarly, instead of asking your students to perform a specific matrix or Möbius transformation on a particular figure, ask them which transformations of a given type could send a figure of one specified type to a figure of another specified type (for example send a square to a rectangle, parallelogram, or trapezium).

Assessment is usually reserved for end-of-topic tasks which call upon the student to use standard techniques in familiar contexts (state a definition, prove a theorem from the course, perform a technique), and then perhaps to extend that or to use it in a less familiar context. Formative assessment (mentioned above) gives students an opportunity to get started and then to seek some guidance on what they are doing and on their approach. Summative assessment (for grades) is meant to sum up what a student can do, but very often results in showing up what a student cannot do.

Drawing on tactics suggested earlier in this book, here are some ideas for types of tasks that could be used equally well for study and for assessment, though sometimes with slight modifications.

Mode of Interaction	*Types of Tasks*
Listening to Exposition	Reconstructing ideas: Write down a mathematically sensible sentence which uses the words ... and .... Write down a few sentences that describe what is going on in the topic of ... and use the following technical terms .... Précis or summarise theorem ... in your own words. Write a note as if to an intelligent but unmathematical relative which explains how to carry out the technique ... or which explains theorem ..., or the definition of a ....
Worked Examples	Construct an example of a ... which satisfies condition ... but not condition .... Given a worked example (say of an integral done from first principles by a regular partition), what is exemplary about it? Identify an example from the course which meets the conditions of theorem ...; construct a more complicated example than those given in the course which meets (or fails to meet) the conditions of theorem .... Possible examples are: a function that is not differentiable at a point; a function that is differentiable everywhere except at five points, or at a countable number of points.
Exploring Ramifications and Interconnections	Carefully designed tasks that are likely to bring students up against significant ideas in a topic. (See Appendix A for examples.)
Exercises	Do as many of these exercises as you feel you need to in order to be able to say in your own words how to do a question of this type, or how to recognise a question of this type. Make up (and do) an easy, a hard, and a general question of this type. Make up (and do) a question that shows me that you know how to do questions of this type. Make up (and do) a question that will challenge students on the same course next year. Write out instructions for someone who does not fully understand this course as to how to do a question of this type.
Historical Investigation	Find out about particular mathematicians, or particular applications connected with specific examples, theorems or techniques by searching the web. The results can then be assembled into a single display or website (see Chapter 6).

See Appendix A for a collection of tasks associated with the convergence of series of positive terms.

Consequences of Assessment

One of the unintended side-effects of any type of testing, whether diagnostic, formative, or summative, is that students define what they are supposed to be able to do by what has been tested for in the past. In our own student days, many of us made use of past exam papers in order to spot likely questions and generally get a flavour of what sorts of things we would be tested on. The more formative and summative testing you do during the term, the stronger will be this effect. Thus it is highly unreasonable to put questions in a final examination that are very different from what was set during the term. This observation lies behind the tactic of getting students to generate their own questions that generalise or typify types of problems: it helps them to prepare for recognising it when they encounter it in an examination (see *Tactic: Student Generated Exercises*, p16).

Although it is usual to think of lectures and tutorials as teaching, it is also useful to think of your course as consisting of the tasks that you set your students, with the lectures, tutorials, and texts as supports and resources. This may be closer to the perspective of many of your students than you like to think. For example, the Schaum Outline series of texts aims to give students exactly what they want and what they need. Some consequences of this view are that:

- the content that you think is important has to be tested, and what is not tested may be seen as unimportant;
- the processes that you want your students to use, the ways of thinking that you want them to experience, have to emerge in the tasks;
- the mathematical themes that you want them to encounter have to play a part in the tasks;
- the mathematical powers that you want them to develop have to emerge or be useful in the tasks you set, not just in the lectures and tutorials.

You may want to go as far as to give marks for appropriate references to the use of processes, themes, and powers (see Chapter 5).

Forms in Mathematics

When as mathematicians we find ourselves trying to make sense of a journal article, a text, or exploring a new idea, we have a range of questions that we usually ask ourselves. If students are not exposed to that same rich range of question types, they are likely to form the impression that mathematics is cut and dried, a closed and completed book, rather than a dynamic and organically developing domain of exploration.

Zygfryd Dyrszlag published a list (reprinted in translation in Watson and Mason, 1998) of questions that he uses with students to test out the quality and nature of their understanding. In making sense of Dyrszlag's list, we identified different elements (list A) and different types of mathematical thinking activities (list B).

List A: Mathematical Forms	
• definitions	• facts
• properties	• theorems
• examples	• counterexamples
• techniques	• instructions
• conjectures	• problems
• representation	• notations
• symbolisation	• explanations
• justifications	• proofs
• reasoning	• links
• relationships	• connections

List B: Mathematical Thinking Activities	
• exemplifying	• specialising
• completing	• deleting
• correcting	• comparing
• sorting	• organising
• changing	• varying
• reversing	• altering
• generalising	• conjecturing
• explaining	• justifying
• verifying	• convincing
• refuting	• depicting

The elements in list A can be thought of as mathematical forms, distinguishable in that the style and purpose of the language tends to differ between them, often markedly. Students may not readily distinguish between them, even though an experienced mathematician exposed to extracts from a text could probably identify most of them without any contextual clues.

The elements in list B can also be thought of as powers which students possess, and which they may well have demonstrated in other contexts. The pedagogic question is whether students are being called upon to employ those powers in sessions and when doing tasks, or whether most of the thinking activities have been done for them through the selection of exercises and the form of the exposition.

The two lists provided us with the rows and columns of a matrix which, after some condensation, produced a collection of general question types.

Exemplifying, Specialising	*Completing, Deleting, Correcting*	*Comparing, Sorting, Organising*
Give me one or more examples of .... Describe, demonstrate, tell, show, choose, draw, find, or locate an example of .... Is ... an example of ...? What makes ... an example? Find a counter-example of .... Are there any special examples of ...?	What *must* be {added / removed / altered} in order to {allow / ensure / contradict} ...? What *can* be {added / removed / altered} without affecting ...? Tell me what is wrong with .... What needs to be changed so that ...?	What is the same and what is different between ...? Sort or organise the following according to .... Is it or is it not ...?

Changing, Varying, Reversing, Altering	*Generalising, Conjecturing*	*Explaining, Justifying, Verifying, Convincing, Refuting*
Alter an aspect of ... to see the effect. What if ...? If this is the answer to a similar question, what was the question? Do ... in two (or more) ways. What is the quickest, easiest, ... way? Change ... in response to imposed constraints.	Of what is this a special case? What happens in general? Is ... true always, sometimes, or never? Describe all possible ... as succinctly as you can. What can change and what has to stay the same so that ... is still true?	Explain why .... Give a reason ... (using or not using ...). How can we be sure that ...? Tell me what is wrong with .... Is it ever false that ...? (Is it always true that...?) How is ... used in ...? Explain the role or use of .... Convince me that ....

Here are some examples of how these questions can be used in practice.

- *What* must *be added to, removed, or altered in order to admit ... as well, or to force ...* (What must be altered in order to convert a definite integral into the area between the graph of the function and the x-axis?)
- *What* can *be added, removed, or altered without affecting ...* (What features of a convergent series can be altered without affecting its convergence or its limit?)
- *Tell me what is wrong with ...* (What is wrong with the reasoning that $\int_0^2 x(2-x)\,dx = 0$ because when you evaluate the function at the limits you get 0 in both cases?)
- *What needs to be changed so that ...?* (What needs to be changed so that $3x^2+2x+1$ has real roots, or so that it has a positive maximum at $x=-1/3$?)

Some questions are directed at altering an aspect to see what effect it has.

- *What if ...?* (What happens to the roots of a quadratic if the coefficient of x changes at a constant rate?)
- *If this is the answer to a similar question, what was the question?* (If the derivative of a function is the function itself, what was the function? What if the derivative was the inverse of the function? If the derivative of $F(x)$ is $f(x)$, what function has $f(x)+2x$ as its derivative?)
- *Of what is this a special case? What happens in general?* (Rolle's theorem, Taylor's theorem)
- *Is ... true always, sometimes, or never?* (Is it always the case that if a function has a derivative at a point it is continuous at all points nearby? Is it always the case that the centroid of a figure lies inside the figure? Is it always the case that $xf(x)$ is continuous at $x=0$?)

- *Describe all possible … as succinctly as possible.* (Describe all possible polynomials through $(1,0)$, $(2,0)$, and $(3,0)$ as succinctly as possible.)
- *Give me one or more examples of …* (Give me an example of a function which is continuous at a point but fails to have a derivative there. Give me an example of a function which fails to be Lipschitz.)
- *Is … an example of …?* (Is this matrix an example of a rotation matrix?)
- *What is exemplary about …?* (What is exemplary about the positive rationals as an example of a group? What is exemplary about $\int x \sin x \, dx$ as an example of an integral achieved by using integration by parts?)

The point about these questions is that they reflect the thinking of mathematicians probing a colleague who is stuck on a problem or idea. We found that these both captured our thinking when preparing and conducting sessions, and reminded us of forms that we had a tendency to ignore. As a result, our own practices expanded.

Learning Outcomes

It is becoming fashionable again to specify objectives for courses in the form of skills acquired as *learning outcomes*: 'After completing this topic, you will be able to …'. Such overt and specific objectives were largely abandoned in the 1970s as being either too general to be informative, or too precise to be comprehensible, but they are being brought back under the labels of transparency, accountability, benchmarks, and standards. There are, however, severe difficulties which have not gone away, because they are part of an endemic tension: the more explicitly you specify the behaviour you seek, the easier it is for students to display that behaviour without generating it from understanding. This is called the *didactic tension* (see *Issue: Implications of Student and Tutor Expectations*, p166) and it pervades teaching and learning, in all subjects, not just in mathematics.

Most attempts to delineate objectives are cast in the form, 'At the end of … the students will be able to …', or, ' … will develop the skill to …'. However, it is rarely indicated whether that 'ability' or skill use depends on receiving specific cues, or whether the students are expected to apply the skills in completely novel and unfamiliar contexts without customary cues. Nor is it stressed that what is learned depends significantly on how the student sets about their studies.

To assert that students will develop some specific skill simply by attending lectures and tutorials, reading texts and handing in assignments, is unreasonable because students can usually find some way to meet all the requirements without actually engaging with the ideas involved. It requires great confidence as a teacher to claim that all students who 'complete' a course will actually learn something specified in advance. Cause –and effect is not the main mechanism behind learning, since, after all:

- just because I possess a text it does not follow that I know what is in it;
- just because I have read through a text it does not follow that I understand the content or can do relevant exercises or problems;
- just because I have worked through the worked examples it does not follow that I can do them by myself or on another occasion, or that I understand the concepts;
- just because I can repeat definitions, it does not follow that I understand or appreciate their significance;
- just because I have done all the problems and exercises myself it does not follow that I will be able to do new problems at a later time under different conditions.

Just because I 'study' a course, it does not follow that I learn what the authors expect me to learn, or that I learn at all or to the extent that the authors intend. A great deal depends on what the students actually do: how they engage, what they do when they are 'working through' a text or set of problems.

An alternative formulation of *learning outcomes* places more responsibility where it belongs, namely on the student, acknowledging that different student actions will produce different kinds of learning.

What you *can* do, however, is to indicate what skills students are expected to develop as part of a more complex collection of expectations. For example, it is unreasonable to expect that students who have just encountered a new idea will be fully confident and conversant with it, so the course and topic expectations can be cast in the following language:

- see a specified idea for the first time;
- gain experience of specified ideas already met previously;
- achieve mastery (facility) with specified ideas and techniques;
- on completion, students will have had opportunities to explore, encounter, and gain facility with specified ideas.

The *opportunity* curriculum is perhaps more useful as a way of thinking than the *guarantee* curriculum. At the heart of this alternative is a framework which distinguishes three levels of engagement with content: *See – Experience – Master* (see p189).

Alternative Formulation

I suggest three forms of learning outcomes, encounters, experiences and competencies, which can be applied at the level of a week's work, the course as a whole, and the degree level. They are cast in terms of ways of using course materials, lectures, and tutorials as a resource to be drawn upon.

The important thing is that students are told what sorts of actions they have to take in order to achieve corresponding learning outcomes.

Encounters

- If you read through the text, (watch the videos,) go to tutorials, work through the worked examples, and try some of the problems; and
- if you use worked examples as templates for doing assignments and for practising on exercises, and if you use the texts largely as a resource to enable you to complete assignments;

then you will encounter the techniques and concepts. You will be aiming to pass the tests of facility and understanding.

Experiences

- If you take opportunities to formulate for yourself what characterises a question of a particular type or which is amenable to a particular technique;
- if you work *on* as well as *through* exercises by telling yourself how to do a question of that type, perhaps checking with other students and/or your tutor;
- if, when you encounter a new or recently introduced technical term you make connections with your last encounter, asking yourself what is similar and what different in the way it is used;
- if you make a note of the concepts that you find elusive and make an opportunity to ask someone (a more experienced student, a tutor, or your lecturer) for elucidation;

then you will be developing a firm foundation for subsequent mastery of the techniques, concepts, principles and ideas. You will be aiming to do fairly well on the tests of facility and understanding.

Competencies

- If you rehearse mentally the enunciation of results and the definitions of important terms while you are walking or 'waiting for a bus';
- if you use technical terms to express in your own words what the theorems and concepts are about;
- if you draw your own diagrams, pictures or notes and talk to yourself as you do so, perhaps checking these with other students and/or your tutor;
- if you construct your own examples to illustrate concepts, theorems, or techniques;
- if you tell yourself a story about the essence of a unit of work, perhaps checking this against other students and/or your tutor;
- if you construct problems like those you have encountered, perhaps generalising where possible;
- if you tell yourself how to do questions of a given type that appear in the course;

then you will be on the way to achieving competency in the use of those techniques and skills. You can expect to do well on the tests of facility and understanding.

Learning is a complex matter. Skills are manifested in students' behaviour, but that behaviour needs to be generated by understanding, not by mere repetition of standard techniques on routine questions. Education is about awareness as well as about behaviour, about thinking and imagining and appreciating and connecting as well as about using standard techniques.

If you assign your students a collection of tasks to do, you need to be confident that simply completing those tasks will actually have the desired effect, or else you will need some more comprehensive statements such as the ones suggested above. We know that it is often possible to 'get the answers' without having much idea of what you are doing or why. It is all bound up in an implicit contract that quickly develops between teacher and student: if the student completes the tasks set, then the student will learn. (See *Issue: Implications of Tutor and Student Expectations*, p166). If, on the other hand, you tell your students what they are supposed to learn during a particular period, for example telling them:

- that they will have seen certain ideas for the first time;
- that they will have experienced certain other ideas a number of times so that they will begin to appreciate their meaning, significance, and intricacy;
- that they will have opportunity to gain facility in certain other ideas so that they will be able to reconstruct techniques for themselves or otherwise have internalised them;

then you will provide a sensible basis for evaluating their progress and engaging with the material.

Expecting specific tasks to have given learning outcomes is highly unrealistic, even if you include in the task specific stimuli for reflection along the ideas suggested earlier ('Make up your own example.', 'How would you handle this type?', 'What characterises a type?', and so on). Indeed, the more specific you are about what a particular task is meant to achieve, the less likely your students are to attain it, because the effective part of a task is 'inner', not 'outer'.

Inner and Outer Tasks

Dick Tahta (1981) suggested that effective tasks actually have more than one aspect, and that it is this quality of multiplicity which makes them effective, not some honed property which forces all students to learn. There is, of course, the overt specification of what one is supposed to do or find (for example find the answer, apply a technique, solve a problem, explore a conjecture or generalisation, or find a proof or a counter-example). However, in any task worth undertaking there are also inner, unspoken aspects; aspects which, if spoken about and made overt at the beginning, would make it more difficult, not less, for the student to gain useful experience.

At the simplest level, the point of doing a collection of exercises is not to get the answers but to reconstruct or re-generalise a theorem, type of problem, or technique from the particular cases. (See *Tactics: Making an Example be Exemplary*, p29; *Student Generated Exercises*, p16; *Advising Students How to Make the Most of a Worked Example*, p97.) A task can also bring students into contact with mathematical thinking activities, as in list B above, as well as with major mathematical themes (see *Themes*, p192).

As already noted, a task can be useful if it brings students up against some awkward cases or unexpected difficulties, such as the impossibility of a single meaning for $0/0$ or 0^0. However, if you tell your students that they are going to experience the impossibility of a universal definition of one of these, they may not commit themselves to trying to find a value. They are unlikely to engage in committed debate with colleagues who advocate other meanings, and hence they may not actually experience the extent of the problem. You may feel that you can successfully present two or more different approaches (for example by consider the sequences 0^{a_n} and $a_n^{\ 0}$ with a_n approaching 0, or consider the limits o $\frac{x-2y}{2x+y}$ as (x, y) approaches the origin along different curves), and you may capture the attention of some students, but perhaps not all. It may all pass them by in a fog because the initiative is all coming from you. Of course, there are many intermediate positions, where you make suggestions and guide their exploration. The important question to ask yourself with any task is whether your students are likely to encounter the important mathematical ideas and whether they will be called upon to exercise their powers to think mathematically.

There are other inner aspects as well. For example, most tasks, whether sets of exercises or explorations, serve to reveal student propensities.

- Are they in the habit of diving in immediately to the first part of the first question, or do they look over the whole set, or consider other possible approaches before starting work?
- Do they give up as soon as they get stuck?
- Do they take one look at certain problems and decide that they are unlikely to succeed?
- How quickly do they look for help?
- Do they have strategies for dealing with being stuck?
- Are they satisfied with an answer to a particular problem, or do they try to generalise?
- ... and so on.

Good tasks can be used to reveal such propensities, so that your students can begin to develop strategies or habits that enable them not to be slaves to them in the future. Tasks can also be used to bring students up against major mathematical themes (see *Themes*, p192), or to make connections with other mathematical topics. Again, if these are made overt at the beginning, then the encounter is liable not to be a personal 'discovery', but rather part of the wallpaper that constitutes the mass of information you have given them.

Tactic: Locating Inner Aspects of Tasks

Tactic: When setting or constructing a task for your students, make a note of what you expect they will encounter in the way of wrinkles or features of the specific ideas, mathematical themes and thinking processes, and personal propensities. Then ask them later whether your conjectures have been substantiated. They may not be very articulate at first, and may not even know what you are talking about. However, with a little experience, and a certain amount of pointing and probing from you, they will begin to shift from the mere doing of tasks, to learning something from doing them.

Challenge and Routine

There is always a tension between setting tasks that are too challenging, and ones that are too routine. Everyone wants their students to be successful, since confidence comes from a sense of having successfully used your own powers of thinking, knowing you have thought your way through a difficult problem (not just found worked examples that matched the assignment tasks), and confidence can be in short supply on entry to the new world of tertiary education. However, setting tasks that can be accomplished without thought, which have little or no *inner content*, is less than unproductive. It can lull some students into a false sense of security while boring others to the point of withdrawing from the course.

Tactic: Making a Task Appropriately Challenging

Tactic: Every task can be made suitably challenging simply by including a component that stimulates your students to shift their attention away from doing and onto what they are doing, why and when it works, and what it illustrates. At various times, try inviting students to:

- explore variations, extensions, and generalisations;
- construct an easy, a hard, and a general problem of the same type (see *Tactic: Student Generated Exercises*, p16);
- describe what sorts of problems the technique will resolve, how one might recognise a problem of that type, and what constitutes being of that type;
- decide what mathematical themes they have encountered while working on the problems;
- recall what mathematical strategies or powers of thinking they used;
- reflect upon what they have learned from working on the problems.

Student Propensities

As already indicated, mathematical tasks can serve an additional purpose by bringing students face-to-face with their own propensities, which is the first stage in learning to overcome habits that are no longer useful. For example, recognising that one has a tendency to follow the first idea that comes to mind is the first step in learning to pause, to savour the pleasure of having discovered a possible line of attack, and check to see if any other possibilities are open before embarking on the first. Similarly, instead of just giving up when you get stuck, one can learn to invoke alternative behaviour, such as at least making a conjecture and assembling evidence, making a list of all the things one knows, and re-describing the things one wants to find out in one's own words (see Mason *et al.,* 1982).. However, you have to become aware of your propensities before you can hope to do anything about them. Again, rarely using diagrams may at times be a strength, but at other times it can be a decided disadvantage. Being satisfied with a specific answer to a specific question may get the homework tasks done quickly, but the effects may not be long lasting, whereas looking for alternative approaches increases your flexibility in the future. Again, it is necessary to become aware of your habits in order to circumvent those that are ineffective.

Different students display different propensities at different times, so it is dangerous to label students as bright or dull, lazy or keen, diligent or negligent. Each of these descriptors may apply to certain behaviours, but that student may display different behaviours according to the circumstances. Thus some students may not show their reactions in ways that you can easily read, while others may respond overtly. To gauge a student by your reading of their behaviour does a disservice to them and to you.

Labelling behaviour (for example, performance on tests or articulacy in class) is one thing; labelling the student is quite another. If you form a strong impression about a student's strengths and weaknesses, you are very likely to act in such a way as to reinforce those. A student who is sensitive to your moods, or desirous of attention and feedback from you, is likely to act in ways that conform with your assumptions. Thus the two of you conspire unwittingly to reinforce the labels formed. Once the student is labelled, the tutor is reinforced in acting as if the label applies to that student all of the time, and the student responds by acting the way it seems the tutor expects. Consequently a cycle, usually but not always negative, is formed. By labelling only the behaviour, the student is in a position to offer different behaviour, and the tutor is in a position to be sensitive to different behaviour.

Some students prefer to mull things over, preferably by themselves, while others are willing to speak out in a small groups, and others do not mind what size group they are in. Encouraging students to be active in their learning is not the same thing as demanding that everyone act in one particular way. Room must be left for different ways of working.

Tactic: Labelling Behaviour *Not* People

Tactic: Whenever you find yourself tempted to label a student, summon up an image of the student's behaviour, and attach the label to that behaviour. When you find yourself tempted to employ a label, try to focus on the behaviour. If it is something that you find uncongenial, then speak to the student about the behaviour as if it were separate from the student.

Hares and Tortoises

Catering for different aptitudes and learning styles (hares and tortoises) is an age-old problem. Indeed, whenever two or more students are gathered together, there are bound to be different styles in evidence. Hares give the appearance of grasping quickly; they make leaps, but sometimes they have only a superficial grasp of the material. Tortoises may seem slow and tortuous, but they often master detail. Hares may overlook what tortoises find out and master, while tortoises may be so caught up in details that they fail to reach an overview of what they have been doing. This is why it is worthwhile setting different kinds of tasks, stimulating your students to work against their propensities to be hares and tortoises.

It may be tempting to label one student *diligent* and another *tardy*, or *lazy*, on the basis of a few interactions, but it is well to remember that your students' behaviour is influenced by many factors. It is the behaviour that strikes you, but the person manifesting the behaviour is probably much more complex than any single label can capture. Different students perform differently on different tasks and at different times. Some students confidently manipulate algebra while others make mistakes like those suggested in the first chapter. The phrase *mixed ability* is commonly used, but makes the mistake of using *ability* as if it is a quality of the student rather than an attribute of some particular observed behaviour. You can only decide that someone is able to do something if they manage to do it. Not doing it is only absence of evidence, and absence of evidence is not evidence of absence. For example, when a student does not answer a specific question, it may not be because they do not know how to use the relevant technique, but because they do not recognise the relevance of that technique. There are legions of examples of students who score poorly on tests but who go on to become experts in their chosen domain, as well as students who do well on tests but do not go on to do creative mathematics.

The imprecise notion of *ability* is a good example of a label which assumes that certain behaviours (responding articulately in class, handing in good assignments) are indicators of a more general quality of the student (high ability). Such a label can all too readily become attached to the student rather than to the behaviour.

Tactic: Hares and Tortoises

Tactic: Try to include in each assignment some questions that stimulate 'hares' to slow down and attend to detail (perhaps by including a technically tricky example), and tortoises to get an overview (through some sort of generalisation task which involves constructing a convincing argument) as well as providing challenging tasks which extend their intuition and go beyond the routine techniques needed for the examination.

Some students are satisfied if they get answers to the tasks they are set; others are only satisfied if they think they could answer similar tasks in the future. Both groups would benefit from tasks that focus them on getting an overview.

Reproducing Oneself

There is a natural tendency for each teacher to respond to and reinforce the propensities of those students who respond to them. This often means that students who are somewhat like them in approach or manner are the ones who get most of the attention. Consequently, it is easy to overlook or dismiss those who do not respond or who work in different ways. The result is that we pick out 'people like us' and overlook the others. Thus it is easy to end up classifying as bright (or not so bright) people who do (or do not) reflect our own ways of thinking. If the tutor's attention is on getting their students to obtain correct answers to exercises and assignments, then their emphasis is likely to be on specific behaviour. If, however, the tutor's attention is on mathematical thinking processes and themes, on trying to get their students to use their own powers to think mathematically, and on impelling their students to generate appropriate behaviour from an appreciation of underlying structures, meaning, and connections, then there is a greater chance that they will support their students' individuality and personal strengths.

Boundary Examples

If students are only offered well-behaved examples, or examples which have additional but irrelevant features, then the reason for careful statements of conditions to a theorem or definition may pass them by and they may well develop the idea that it is possible to have ambiguous or undecided cases. For example, MacHale (1980) pointed out that although students are repeatedly shown the function $x \to |x|$ on the reals as a continuous function that is not everywhere differentiable, they often still treat it as an isolated example, and he lists a number of other similar dismissals on the part of students. If students are invited to use this to construct their own examples of functions not differentiable at $1, 2, 3, \ldots$, or countably many, points, they might begin to see that it is not a single isolated example. If they are asked to decide what is the same and what different about $x \to |x|$ and the function $x \to x$ for $x \leq 1$ and $2x-1$ for $x \geq 1$, they might begin to see an even wider range of possibilities opening up. They could even be asked to construct functions that are continuous and have slope n on the interval $[1/n+1, 1/n)$, or which are n but not $n+1$ times differentiable at $x = n$, for all n, and infinitely differentiable elsewhere.

There is lively debate among lecturers as to whether it helps to expose students to 'pathological' examples. On the one hand, students who struggle to understand an example may not appreciate its significance. On the other hand, if students do not experience the tug at their intuition that comes from discovering that a formal definition has admitted a whole class of objects which they had not even considered previously, then the force of the theorems proved using the definition may go right over their heads. If your only examples are polynomials or their equivalent in other domains, then the theorems seem largely obvious and the fuss made about proofs can remain a complete mystery.

If, for instance, students have only experienced monotonically increasing or decreasing sequences, they may logically believe that no sequence can attain its limit. On the other hand, if they have also encountered sequences such as $\{1, 1, 1, 1, \ldots\}$, and appreciated the generic nature of such an example, they may be more aware of the infinitely many ways a sequence can incorporate a constant subsequence, or take its 'limiting value' in the sequence. With such experience, students are less likely to focus on the irrelevant feature of increase or decrease. Note that we are not talking about specially constructed pathological examples here, but a class of rather ordinary examples that may extend the boundaries of students' understanding. How can students be helped to appreciate the importance of such examples? How can they be helped to focus attention on relevant features?

Tactic: Boundary Examples (again)

Tactic: Start by constructing examples showing that each of the conditions in a theorem is necessary, but over time make this less and less direct, and more and more subtle or indirect, until your students are doing it for themselves. You can set similar tasks for homework, such as giving them some examples and asking why they do not contradict the theorem. Later, you can give them similar tasks sporadically as a reminder that they could be doing this for themselves.

It is well known that students tend to retain their naïve intuitions about continuity, limits, and forces, and that they struggle, for example, with the notion of elements of sets such as groups as objects not operations despite being introduced to formal definitions, and even appearing to handle them competently. Indeed, Fischbein (1987) believed that intuition is never displaced, merely overlaid. Even if one does not entirely agree with him, the fact remains that students' sense of, or concept image for, a technical term is often an interwoven mixture of formal definitions and intuitions.

Conjecture

If you cannot construct boundary examples for a theorem or a technique, then you do not fully appreciate or understand it.

Note that the claim is *not* that construction of examples constitutes understanding, only that constructing your own examples, stimulated by a task that prompts you to explore a range of possibilities, makes a very useful contribution to achieving familiarity and competence. Students can be prompted to make deeper sense of theorems and techniques only if they themselves have constructed examples and non-examples.

Constructing Examples that Meet Given Constraints

A common approach to counter-examples is to provide students with examples, as in the following.

- Draw a sketch of a continuous real-valued function with domain $[0,1]$ and co-domain $[0,1]$ that is not differentiable at a point in the interior of the interval.
- Write down a polynomial that passes through the points $(1,1)$, $(2,3)$, and $(3,5)$.
- Write down an infinite sequence that diverges even more slowly than $\sum_{1}^{\infty}\frac{1}{n}$.

These are standard objects that are often offered to students in courses. The trouble is that they are usually provided by the teacher rather than the students, so many students do not spend time working on the example to see why it has the claimed properties, nor for which conjectures it provides a counter-example and why. Each is both an opportunity to get the students to construct examples themselves, and an opportunity to generalise.

- Which subsets of the reals can be the set of points at which a function is not differentiable? Build up a class of such functions that shows the possibilities.
- Find a general method for writing down the most general polynomials that pass through a finite set of specified points, such as $(1,2)$, $(3,5)$, $(5,8)$, $(8,13)$, ... Build up a general procedure for answering such questions.
- Build up a collection of different ways to modify a sequence so that it converges or diverges more quickly (or more slowly) than a given sequence.

Here there are two important features: getting students to construct their own examples, so that they have to check conditions and at the same time develop fluency, and getting students to contemplate the general class of such examples. The next section offers a particularly good way to promote this.

A powerful version of example-construction arises when constraints can be meaningfully compounded, so that examples are required which meet all but one of the constraints. It is particularly informative to accumulate the constraints in order, and then to work backwards to find boundary examples, as in the following example.

- Draw a sketch of a real-valued function with domain $[0,1]$, and co-domain $[0,1]$.
- Draw a sketch of a real-valued function with domain $[0,1]$, and co-domain $[0,1]$,that is also continuous.

- Draw a sketch of a real-valued function with domain $[0,1]$, and co-domain $[0,1]$, that is also continuous and which also has a maximum on $[0,1]$ at one end-point.
- Draw a sketch of a real-valued function with domain $[0,1]$, and co-domain $[0,1]$, that is also continuous, which has a maximum on $[0,1]$ at one end-point and which has a minimum on $[0,1]$ at the other end-point.
- Draw a sketch of a real-valued function with domain $[0,1]$, and co-domain $[0,1]$, that is also continuous, which has a maximum on $[0,1]$ at one end-point, a minimum on $[0,1]$ at the other end-point, and which has a local maximum and a local minimum in the interior of the interval.

Now comes the interesting part. Work your way back *up* the list, making sure that, at each stage, your example *does not* satisfy the next (more demanding) condition. Thus your first example must be a discontinuous function, while your last but one must have its extrema at the endpoints but not have local maxima or minima in the interior of the interval.

The virtue of an exercise like this is that students come to appreciate the range of possibilities from which they can select an example. 'Example' can be interpreted to mean 'a special example' or 'an obvious example'. Indeed, one variant is to ask students at each stage to construct a particular example, then a peculiar example (for example one that no-one else in the class is likely to think of), and then (if appropriate) a general, or at least maximally general example. The effect of seeking a 'peculiar' example is to open up for consideration the range of possibilities, the scope of the general case (Bills, 1996). (See *Tactic: Particular – Peculiar – General,* p88.)

In constructing a boundary example, students are forced to extend their example-space in order to complete the task. So one effect is that students become more aware of the range of possibilities from which they are choosing when they select an example, and this is a precursor to their expressing the full generality of the concept.

Other Examples

- Write down a group of order 8; which is also Abelian; which is the direct product of two non-trivial subgroups; which has an element of order 4. Working backwards up the list, the first challenge is to find a group of order 8 that is a direct product but with no element of order 4, while the last is to write down a group of order 8 which is not Abelian.
- Write down a sequence of positive reals; which converges to 2; which takes on the value 2 infinitely often; for which infinitely many values are not 2; for which for any positive integer n there is a term beyond the nth which is less than $2-1/n$.

- Write down a linear transformation $R^3 \to R^3$; which is singular; which has $(0,0,1)$ in the kernel; which also sends $(1,0,0) \to (1,0,0)$; which also has $(0,1,0)$ has an eigenvector; with eigenvalue 3.

Again, most of the force of the work comes when working back up the list seeking boundary examples.

Of course it would be possible to discuss and describe the most general object possible at each stage. It would also be desirable sometimes to introduce constraints that are impossible to fulfil. In the case of the function sketching example, discussion of exactly what 'local maximum' means is likely to occur (do you accept a function which is constant at the maximum value near the endpoint as having a local maximum?). Provoking students to discuss extreme or special cases such as empty intersections and unions, circles of zero or infinite radius, or triangles with an angle of zero contributes to their impression of the range of possibilities covered by the conditions.

Final Comments

Using a sequence of increasing constraints has the added virtue that it displays mathematical objects as the result of *freedom* (to choose any example from a wide range of possibilities) within *constraints*. For example, solving linear equations is really an exercise in seeking the points that started as being general or free, but now have to satisfy a number of constraints. Finding a method for making a geometrical construction can similarly be seen as starting with initial freedom and then adding constraints.

Becoming aware of a range of mathematical objects enhances students' sense of generality, as well as preparing them to recognise which technique might be useful in a given situation. Seeing 'solving a problem' as seeking a general object that has had constraints placed upon it can totally alter the a student's perspective on the task.

Many students do not appreciate the range of objects that are permitted by a theorem or a technique. They see techniques as tools, but they are usually rather hazy about the nature of the mathematical objects that they seek as solutions to their problems. Yet most theorems can be seen as a description of something that is *invariant amid change*, with the conditions of the theorem stating the range of change that is permitted. However, if students have not tried to construct examples for themselves, have not probed the role of various conditions in making a theorem or technique work, then they are unlikely to use it appropriately, and probably unlikely to think of using it at all!

Finally, there is the ubiquitous concern about time. The first time you ask your students to construct an example for themselves, most will probably not succeed, at least without a fair amount of time and help. How, then, can all the material be 'covered'? My answer is that, over a period of time working at getting your students to construct examples, they will become much better at it. In the process, they will also become much more proficient in the mathematics involved. Consequently, they will begin to learn more efficiently, and the pace of the course can accelerate (but neither constantly nor uniformly!). In effect, we cannot afford *not* to invest the time needed in order to enable students to appreciate the ideas to which they are being exposed.

Reflection

Task: Reflection

What do now you see as the principal purposes of setting tasks for your students? If there is something that you would like to use or develop, imagine yourself actually doing it, and experience the pleasure of achieving your aim. This is the best way to support yourself in developing your practice of teaching.

5 Marking and Commenting

Introduction

Having discussed how to generate suitable assessment tasks in the preceding chapters, we now turn to the topic of how best to comment on your students' scripts. Remember that each tutor has their own style, so that the same comment coming from two different tutors may seem forced or artificial for one but completely natural for the other. Furthermore, some students may take offence at something that, though meant humorously, could be interpreted as condescending or demeaning, so great care is needed when being light-hearted. Consequently, it is not possible to specify precisely which comments will be helpful. It is essential to find your own style, while at the same time thinking about your students' experience of receiving comments.

Task: Receiving Comments on Work Handed In

Do you remember what you used to do, when you were a student, with work that was marked and handed back to you? Did you work through it again assiduously, or did you glance through it and then put it to one side? How do you feel when a colleague points out an error or disagrees with an argument that you are developing?

Comment: Many students do not look at comments until close to examination time, if ever, unless the comments are focused and pertinent.

The sooner comments are returned, the more likely a student is to pay attention to them. No one likes to be told that they are wrong, so writing comments which students find helpful is a non-trivial activity. The suggestions in this section are based on Davis (1987) and are divided into two subsections: being clear about how marks are allocated, and providing useful feedback to students.

Allocating Marks

Marks without comments are of little value to students, and are likely to discourage them from going over their work and trying to learn from what they have done. Providing a 'perfect solution' which gives no clues as to how the tutor knew what to do at each stage is equally unlikely to help many students, even if they think it will. What they need are comments which help them to see how an expert tackles questions, and some suggestions as to how to do this are made in the rest of this section.

At the very least, indicate where you have followed through despite an initial or on-going error, where you have given a mark for method, for an answer, or for a process, and where you have deducted a mark for an error. The more precisely you can tell your students how their work will be marked, the fairer the marking is likely to seem and the more they will learn from it.

Tactic: Using Group Convenors

One of the best ways for students to learn what a marker expects is to do the marking for themselves.

Tactic: Appoint one person per five or six students as a convenor (preferably with a different convenor each time). The members of the group then submit their homework to the convenor, who marks it and presents a summary of the approaches of their group to everyone. The presentation can either be made just to the group, or to the whole class. (This tactic was suggested by Ed Barbeau.)

This tactic supports the convenor in learning to make sense of, comment on, and judge other people's work, as well as forcing them to extract and abstract the different approaches taken from the technical details.

Tactic: Collaborative Work

Tactic: Ask your students, in groups of five or six, to present their homework to the rest of their group. Either the group agree a common grade, or each member suggests a suitable grade and the various contributions are used to create a consensus submission. Really good expositions can be nominated for full group presentation.

Tactic: Rewarding Awareness of Processes

Right answers are important, but so is the use of the appropriate method. You can support students in choosing an appropriate method, and more generally in thinking mathematically, by using marks at the beginning of a course to reward the explicit use of mathematical thinking processes.

Tactic: At first you can give your students marks for aptly including comments on *what* they are doing and *why* they are doing it, as well as for competently carrying out techniques. Useful comments (for the student as well as for the tutor) include, for example: 'Stuck!'; 'Conjecture!'; 'Specialising'; 'Generalising'; 'Rearranging'; 'Using theorem X'.

At the Open University we used to set an assignment consisting of a report of an exploration from a set starting point. Half the marks were allotted for appropriate identification of mathematical thinking processes, and we found that some students carried on using some of the same words in assignments in later courses. However, other students merely scattered a few words on the page, doing the minimum they felt necessary in order to pass, while some, especially those who found the work relatively easy, simply handed in work without any reflection or commentary. Some students do feel frustrated by this sort of scheme because they feel that they do not need to reflect on their thinking, and that marks are being handed out too freely! They may even have great difficulty in reflecting precisely because they do not (yet) struggle.

However, they will have every opportunity to shine later, and, since everyone gets stuck sometime, they may also learn something about what to do when they do get stuck.

While teaching a masters course in India in the sixties, Bob Burn (private communication) reserved 10% of the marks for questions which his students posed for themselves. They scored 10% for asking at least one good question per course, 5% for bad questions and 0% for no questions. To his astonishment he never had any difficulty distinguishing between good and bad questions. However, he did find that he had to be ready to engage with his students when they asked questions. He kept it up for four years with a small group of 16 students. The driving force was his observation that

The methods of assessment are the clearest way we have of telling students what we regard as important.

There is clearly an element of subjectivity in grading what students write; the important point is that they are writing about what they are doing, not just doing it, and that their comments are reasonably appropriate. You are not judging whether they did actually think the way they say, but whether that thinking is appropriate.

Do not forget that the more you encourage students to write, the more time it is likely to take them and the longer it will take to mark and comment. There may be knock-on effects when judging the appropriate length of assignments, in that the more time it will take to mark them, the harder it will be to get graduate students or other colleagues to help with the marking.

It is of more benefit to students to know what they have to do to get a similar or better mark next time, than it is to know what mark they received this time. Consequently, showing where a mark was deducted is more helpful than just showing the number of marks accumulated. Commenting explicitly that in order to get the same mark next time they would need to do more (for example provide more detailed commentary on what they are doing, or show more details of their calculations or proof steps) is more effective than simply docking marks this time and expecting that to make them learn.

Students need to see what a full solution looks like, at least every so often. You can write these yourself, or use good ones from previous years. It is probably more useful to students to see a fairly good script with your comments on it showing how it could be improved, than to see a perfect script.

See also Angelo and Cross (1993), and Gold (1999).

Feedback Among Markers

While students often report that they feel isolated when doing assignments, markers can have a similar experience, especially when they are inexperienced graduate students earning money by tutoring and marking. To overcome this isolation, and the concern that perhaps others are marking more leniently or more rigorously, there are a number of helpful tactics.

Tactic: Being Clear About the Mark Scheme

It is vital to have a clear and detailed mark scheme, not only when more than one marker is involved, but also if you are marking them all yourself. It is amazingly difficult to maintain uniformity when you mark scripts over a period of time, or when you have a large number of them. In any case, it is essential to have a detailed mark scheme in case students challenge the mark you have given, as well as, in the UK, for inspection purposes by the subject review panel.

Tactic: Specify what a student has to do to gain (or lose) each mark you assign. By all means refine the scheme in the light of experience but, if several markers are involved, they all need to be kept informed.

It is common practice, where more than one marker is involved, to mark several scripts collectively, chosen either to illustrate particular difficulties that may arise or show the work of weaker students.

Tactic: Once you have specified what a student has to do to get the marks, why not remove any specific answers or methods, and then provide students with the mark scheme?

Tactic: Establishing an Atmosphere of Mutual Support

Tactic: Establish an atmosphere among markers such that they are willing to show difficult scripts to each other and to engage in discussion about marking.

This means being interested yourself in how the marking is proceeding, and, even better, marking some scripts yourself. It also means being open to discussion about what different questions require of students, whether the same thing could be tested in other ways, and so on. Such discussion could be invaluable for graduate students for when they find themselves setting and marking scripts in their first lecturing job. However, do not forget that, just as students want to minimise their investment, so script markers want to minimise their effort. Any 'extra' discussion has to have evident value.

Tactic: Exchanging Scripts

Tactic: Arrange for one or two scripts to be marked by everyone, so that some comparisons can be made. Allow markers to see how others mark, especially concerning comments to the student. (This topic is developed in the next section.)

Feedback to Students

Probably the most common feedback comment is, 'See my worked solution'. For a student who is struggling this may be enough, though it is probable that few students spend enough time trying to learn from their mistakes. Many students file comments with the aim of using them to study for the examination, but if they re-worked the problem using your solution as a guide, and then constructed and solved their own version or variant, they would be much more likely to gain from the experience.

Tactic: Focusing on what is mathematical

Marked scripts can be used to indicate where students have gone wrong, but with a little extra effort they can also be used to help students develop effective learning strategies as well as more effective responses to problem assignments. Focusing on what is mathematical about a student's work, even where the computations go awry, can make a huge difference to them, since they are often only too well aware that they are struggling.

Tactic: Be explicit with students about what you are looking for when marking.

Look for opportunities to use comments such as, 'Did you think of trying an example here to see what is going on?' or, 'Very useful to try an example, but it may need to be more complicated in order to reveal what is going on.'. These can be much more helpful than simply pointing out where a computation has gone off the rails.

Tactic: Developing a Language

Writing comments requires a language in which to express your thoughts about learning and doing mathematics. If your language is solely based in the concepts of right and wrong, reflecting a single dimension of mathematical reasoning, then you will fail to make use of other pertinent distinctions. For example, as pointed out in earlier chapters, students do not often understand a new idea immediately; rather, they work away at it, trying to get a sense of what is going on, and trying to articulate that with the help of notes and text. Thus they may have a good sense of what is required by an assignment task, but not be able to write much or anything down, or when they do write something it may be all over the place.

Tactic: Decide whether you want to promulgate a picture of mathematics as worked out in advance by experts or as a disciplined way of thinking open to many different approaches. Then make sure that your language reflects your views.

For example, the word 'hint' suggests a single desired solution, and that as tutor your role is to draw back some of the many veils. By contrast, 'suggestion' suggests openness to different approaches and that the particular offering may not be helpful to everyone.

Being sensitive to your students' struggle to express themselves often includes offering advice such as, 'Try telling your goldfish how you want to solve the problem, then start writing.' As we all know, we often have to express our thinking to someone else before we can write it down successfully. (See *Tactic: Say What You See*, p20, 30, 91; see *Issue: Discussing Mathematics*, p176.)

Tactic: Finding Something Positive to Say

Simply saying, 'See my worked solution' may be just what some students need to clarify or correct an erroneous impression they have formed. However, it is often of little value to a student who is struggling, since they may not be able to work out how you knew to invoke a technique or pick out some detail to elaborate, when they themselves did not think of it.

Tactic: Make your first comment positive, without any sting in the tail. For example, 'You have made a good start, but ...' can be seen by students as a criticism, the positive part being overwhelmed by the negative comment that follows. Instead, find some mathematical thinking to praise, such as effective use of diagrams, good use of an example, or an attempt to generalise. Thank them for their commentaries, as this will encourage them and their commentaries in the future will also help you to work out what they are trying to do!

These suggestions lead to an inevitable tension for the marker between wanting to get the marking done in a reasonable time, and needing to write enough to be helpful. The danger is that a marker will want to write too much, so that the student is put off and overloaded by all the comments.

Tactic: Selecting What to Mark

Tactic: Instead of marking everything in great detail, you can choose to mark some questions in considerable detail, and the others by simply glancing through the work and looking for shape, form, and expected answers. As long as the students do not know in advance which questions will be carefully scrutinised, they are likely to try hard at all of them.

Tactic: Providing Model Solutions

Where many students make the same error, it requires personal discipline to respond as helpfully to the last script as to the first. Just suggesting, 'Try integration by parts.' or, 'Try leaving the polynomials in factored form.' is not terribly enlightening for a struggling student. They need to see (and probably to be shown) why it makes things easier.

Tactic: One approach is to write out a model solution for everyone. However, there is a danger that your attention will be attracted to 'getting the maths right', and away from the specific difficulties your students were having.

What students most need is assistance in working out what to do, often at points that are completely obvious to you. You can help by explaining your thinking, being explicit about the choices you make ('Shall I try a simple example here, or shall I differentiate or do a sketch?').

Tactic: Summarising Your Observations

Tactic: Summarising what your students might usefully think about for the next assignment can be of more benefit to them than reading through your individual comments. On a summary sheet, try to:

- start and finish with an optimistic comment if possible;
- remind them of one or two specific difficulties they encountered and how to correct them;
- remind them of what they need to do to get a similar or better mark next time (do not just tell them to 'get more answers right'!).

Do not forget that your comments on your students' scripts carry a strong message about the nature of mathematics. Some words, used unthinkingly, can have a devastating effect and can make you appear to promote a view that conflicts with your true views. By showing a human face, by being sympathetic and acknowledging that certain ideas are difficult, you can help to retain students in courses.

Tactic: Providing a List of Common Errors or a 'Corrected' Sample Student Argument

The tactic *Using Common Errors* (p10) can be employed by returning to each student a collection of common errors made on the same (or similar) problems. You could also provide a made-up sample student attempt at an argument that you then comment on, augment, and rearrange, so that your students can see the contrast between the sort of thing they might be providing and the sort of thing you are expecting from them. Since you are trying to develop their confidence, these examples work best when they lead to a sense of satisfaction for your students, not just another challenge.

Sample Script A (as submitted by a student)

Question: Show that $x \to x^6+1$ is increasing on $[0,1]$. Note that $x_2^6 - x_1^6 = (x_2 - x_1)(x_2^5 x_1 + x_2^4 x_1^2 + x_2^3 x_1^3 + x_2^2 x_1^4 + x_2^1 x_1^5)$.

Solution: For $x \to x^6+1$ to be increasing on $[0,1]$, $x_1 < x_2$.

Therefore $x_1^6+1 < x_2^6+1$,

$(x_2^6+1)-(x_1^6+1) > 0$.

Therefore $x_2^6 - x_1^6 > 0$.

Given that $x_2^6 - x_1^6$ factors as shown, then

$(x_2 - x_1)(x_2^5 x_1 + x_2^4 x_1^2 + x_2^3 x_1^3 + x_2^2 x_1^4 + x_2^1 x_1^5) > 0$.

For this to be true over the interval $[0,1]$, both these elements have to be positive.

So $(x_2 - x_1) > 0$ and $(x_2^5 x_1 + x_2^4 x_1^2 + x_2^3 x_1^3 + x_2^2 x_1^4 + x_2^1 x_1^5) > 0$.

If $x_1 = 0$, the right hand inequality reduces to x_2^5, which is positive when $x_2 > 0$.

Therefore $x_2^6 - x_1^6 > 0$ is justified and $x \to x^6 + 1$ must be increasing on $[0,1]$.

Comment: Here is a classic example of a proof attempt that has come out backwards. Most of the essential work has been done, but in any proof it is essential to put down an anchor, to plant your feet in the hypotheses, while this student has chosen to anchor themselves in what they were trying to prove.

At first it may be tempting to ask to see the student face to face, but this causes added tensions and may not always be most productive. Just what is it that the student has to have missed out on? Their use of subscripts seems a little odd, but it may be that they are actually struggling to use subscripts in the way that they think they have seen their lecturer use them. Perhaps the difficulty is that they have not established a pattern of starting from what they know and proceeding to what they want. An observation like this might lead you as a lecturer to establish a routine 'rubric' for writing up proofs in notes and lectures, at least for a period of time. Here, you could suggest to the student that they write down what they know, and what they want:

- 'I know that x lies in $[0,1]$, and that $x_2^6 - x_1^6$ factors as a sum of terms.'
- 'I want to show that the function is increasing in $[0,1]$.'

For a tutor who sees immediately what to do, it is not always easy to enter the experience of someone who does not yet know. Here, the important elements seem to be to express 'increasing' in terms of x_2 and x_1 so that factoring can be used. The student may need to write out what 'increasing' means. Ultimately, the student has to learn to take that initial (and often only) creative step:

'Let x_1 and x_2 be in $[0,1]$ with $x_2 > x_1$. Now I want $f(x_2) > f(x_1)$ on $[0,1]$.'

That statement provides a starting point from which to make deductions, and perhaps that is all the student needs in this case.

Sample Script B (as submitted by a student)

Question: Using suitable software, graph the parametric curve $(\cos at + \cos bt, \sin at - \sin bt)$ for various choices of the parameters a and b from the positive integers greater than 1. Make a conjecture concerning the shape of the curves (numbers of loops, numbers of crossings).

Solution: It seemed that the number of loops = number of crossings = $a+b$, until I hit upon $a=4$, $b=2$ which has only three loops and three crossings, and $a=b=4$ which is only a line segment. Further experiments suggested that when $b=a/2$ there are just three loops.

Comment: The student has only partially addressed the problem and has not indicated what examples were tried, but they have indicated a conjecture that was then disproved. Encouraging the student to provide at least a table of experimental results would help both you and them. Suggesting that the student try to find a conjecture that encompasses cases such as $a=4$, $b=2$, may simply reinforce the student's insecurity, since they probably looked for one but did not find it.

What students often do not often realise is that once a conjecture has been made it is valuable to search for non-examples, or to vary the situation looking for properties that the examples chosen so far have in common and which could be modified. For example, suggesting that the student try the case $a=6$, $b=4$ may have the desired effect, but it may not help them to see why that is a useful case to try. If you can become aware of how you know what other cases to try, then you can offer advice to your students. Asking the student to consider what is the same and what is different about the cases $a=4$, $b=2$ and $a=6$, $b=4$ might be a little more helpful, as it might draw attention to common divisors rather than to divisibility.

You could also suggest that the student look for a formula that expresses the sum of two cosines as a product, and the difference of two sines as a product.

When asking students to undertake extended work, it is important to have in mind some exemplars so that you can describe to your students what you are looking for, and even offer one or more examples for them to use. As you build up a collection of responses that display both positive and negative features, you can make these available, perhaps with marker's commentaries. Here is an example of a script that includes a variety of remarks about the exploration, making it relatively easy to follow the line(s) of thought and to see the choices that have been made.

Sample Script C (as submitted by a colleague)

Question: A string is hung over a horizontal cylindrical rod. When one end is supporting m grams and the other end is supporting M grams, the string is on the point of slipping. Find the coefficient of friction between the rod and the string.

Solution: I should start by drawing a diagram.

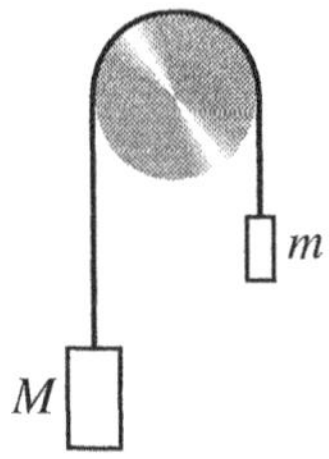

Do I have any instant reactions?

What stops the masses slipping off the cylinder?

What is the radius of the rod?

Is the rod smooth?

How does the coefficient of friction come into this problem?

Have I ever seen anything like this before?

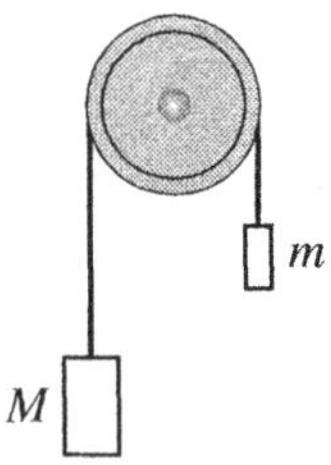

I have seen lots of examples of masses supported by strings over pulleys, but in all the cases I can recall the pulley was frictionless and the string was 'perfect' and 'inextensible'.

The tension in such a string does not vary along its length, and if that were the case in this example then the masses would begin to move. This means that this is not the usual kind of 'string'.

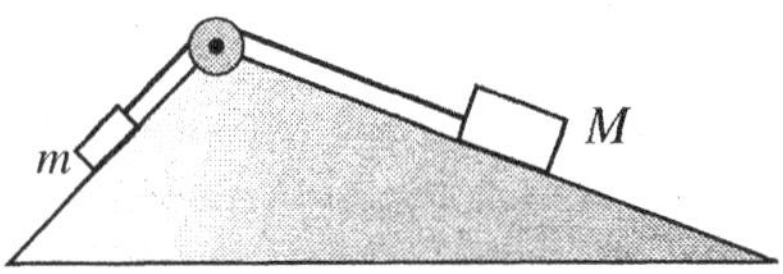

On the other hand, I can recall problems involving inclined planes where two unequal masses were in equilibrium.

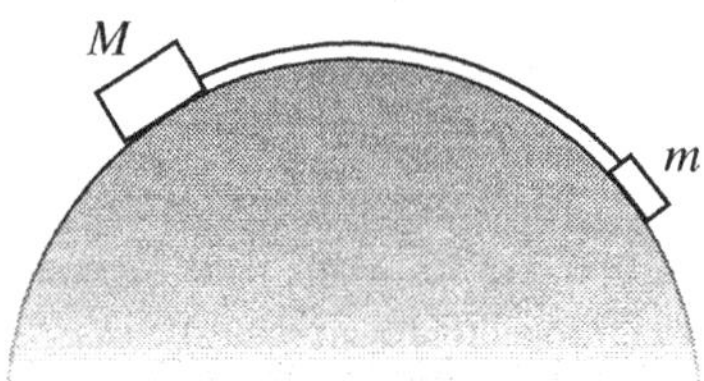

I guess I could arrange for the two masses in the current problem to be in equilibrium if they were arranged asymmetrically on the cylinder.

This looks like a possibility, but I suspect that the problem doesn't mean this, so I'll put this idea to one side.

Do I notice anything unusual about the problem?

The phrase, 'the coefficient of friction between the rod and the string' is very strange. It is also very odd that the radius of the cylinder is not mentioned.

How could the masses be in equilibrium in my original diagram?

The string would need to be rough, and this would mean that the tension would vary along the length of the string. This looks hopeful because we need this to happen if the upward forces on the two masses are to be unequal.

How can I deal with a rough string?

The obvious idea would be to cut it up into lots of small elements, and then to treat each element as a small particle.

Do I need to allow for the mass of the string?

Possibly, but for the moment let's suppose that the string is very light so that I can ignore its mass.

It would probably be a good idea to draw another diagram, and, since the tension depends on position, I'll denote it by $T(\theta)$.

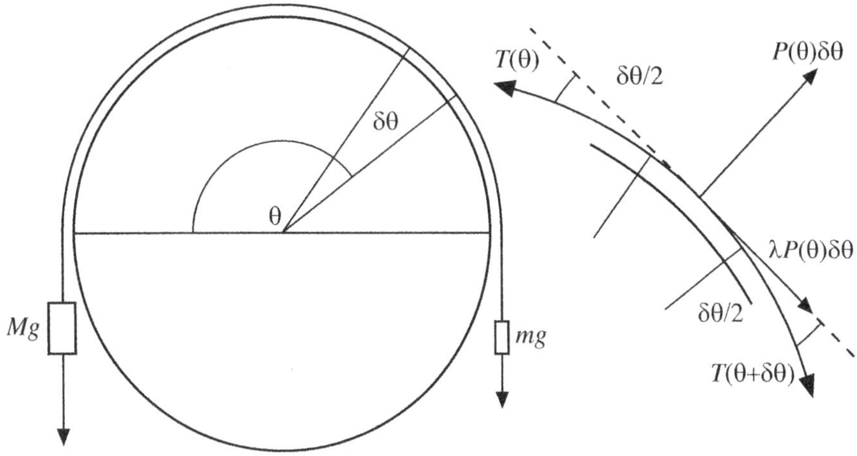

Each small element of string will apply a small force towards the centre of the cylinder, and this reminds me of 'pressure', so it seems sensible to denote this 'force per unit length' as $P(\theta)$.

Can we assume that the string is 'on the point of slipping' at every point along its length?

If the force along an outward normal from the cylinder on a small element of string is $P(\theta)\delta\theta$, then I would like to be able to deduce that the frictional force on this element is $\lambda P(\theta)\delta\theta$ along a tangent (in a direction that opposes the 'motion'). I can only do so if the string is 'about to slip' at every point along its length. I suspect that this has something to do with assuming that the string is inextensible, but I'll put that difficulty to one side for the moment.

The problem is beginning to make more sense now. The phrase 'the coefficient of friction between the rod and the string' makes a lot more sense. Maybe the question intends to say that the string is on the point of slipping at *every* point of contact between the cylinder and the string?

(Added later: if the string is inextensible, and starts to slip somewhere, then it has to slip everywhere, so the force making it move must be greater than the effect of friction at every point.)

Do I have enough information to construct some equations to specify that the system is in equilibrium?

Let's see what happens when I write down the equilibrium conditions (I truly do not know where this will lead).

Resolving along the tangent,

$$T(\theta)\cos(\delta\theta/2) = T(\theta+\delta\theta)\cos(\delta\theta/2) + \lambda P(\theta)\delta\theta,$$

and in the limit this gives

$$T'(\theta) = -\lambda P(\theta).$$

Resolving along the normal,

$$P(\theta)\delta\theta = \left(T(\theta) + T(\theta+\delta\theta)\right)\sin(\delta\theta/2),$$

and in the limit this gives

$$P(\theta) = T(\theta).$$

This starts to look promising because I now have to solve the differential equation

$$T'(\theta) = -\lambda T(\theta)$$

with the appropriate end conditions. This gives $T(\theta) = Mge^{-\lambda\theta}$,

and then $T(\pi) = Mge^{-\lambda\pi} = mg$. From this I get $\lambda = \frac{1}{\pi}\ln\frac{M}{m}$.

Extension: Suppose that the string winds around the cylinder several times, as I have seen on winches, or suppose that it drapes over several cylinders alternately above and below? Assuming that I am only concerned with string against cylinder (not string against string), then all that matters is the length of the contact between string and cylinder. If the string is wound around n full revolutions in addition to the half, then π will be replaced by $(2n+1)\pi$.

The original problem seems to be rather unrealistic, so I am led to an alternative, and I think more realistic, problem.

Question (extended): A long rope is draped over a wooden beam, and a heavy weight is suspended from each end. Is it possible for such a system to remain in equilibrium with neither weight in contact with the ground? Does it matter whether the cross section of the beam is rectangular or circular?

Construct a mathematical model that could be used to predict the behaviour of such a system, and specify the assumptions of your model explicitly. If you need to introduce any physical constants (such as a 'coefficient of friction') suggest how such constants might be measured.

Comment: Having an ideal solution in mind is one thing; having an ideal written down is quite another. However, until one or more are written down it is difficult to be clear about your marking criteria, the features you wish to reward and will be looking for. Once you have decided those, surely the students ought to be let into the secret!

In this case, the student has clearly stated the assumptions they made along the way, clearly indicated where choices had to be made, and offered simplifications and extensions and variations that could also be explored (and in a full project might be expected to be). They could also have checked that the results seemed reasonable.

Feedback from Students

Assessment is primarily for students, but it is only really useful if it also provides information for the lecturer, informing how to change tack if necessary and to modify future presentations and tutorials.

Tactic: Learning from Students

When a significant number of students do not submit what you expect, then there is something to learn. However, how will you know about it if others are doing the marking?

Tactic: Choose one question, that you think is crucial to the next topic, and have your students submit their answers on a separate page. Then either mark all of that question yourself, or get tutors to mark them and then forward to you those that do not get over some threshold so that you can see what is going on.

Students can also be asked to indicate topics, techniques, or examples that they would like more help with, and this can then inform future tutorials or even lectures.

Reflection

Task: Reflection

Is there a script that you marked recently which would have benefited from comments along the lines suggested in this section?

Which of these practices appeal to you with regard to marking scripts? What aspects of them might be difficult to carry through, and why?

Comment: It is vital to be reasonably strict with yourself and consider each of the proposals in turn, much as you would a collection of mathematical conjectures. Imagine yourself using them, and consider what might happen and how you might modify the proposals to suit your own circumstances.

6 Making Use of History

'It seems to me that if one wants to make progress in mathematics, one should study the masters.'

(Niels Henrik Abel)

Introduction

This chapter makes a case for using historical remarks and historical problems, both as part of a mathematics course, and as the basis for a course in the history of mathematics. It then offers suggestions for different ways in which students can be engaged in historical problem-solving without detracting from the mathematical topic.

Why Use History?

Mathematics is often presented as a logical subject that, when laid out clearly step by step, can be readily assimilated by any rational person. Unfortunately, however, things are not quite this simple. Students have personal desires and propensities; their interest is caught by different things at different times, and they find some aspects more difficult than others. Many students become so embroiled in mastering definitions, theorems, and proofs that they develop an all-or-nothing approach: either they understand and can reconstruct every detail, or they do not understand at all. Such students are in danger of losing sight of the human aspect of mathematics, and of the many subtle levels and degrees of its understanding.

One way to provide a human element is to draw upon the history of mathematics. Some students appreciate hearing about the struggles of the people who first formulated important ideas or who were among the first to provide a proof of a significant result, because it shows that mathematics is done by human beings with their own passions and foibles. However, history is more than just a collection of portraits and anecdotes: it is part and parcel of mathematics itself. No matter how the presentation of ideas that were initially difficult is altered and reformulated to make them more accessible, their historical origins and the problems which people were originally trying to solve remain part of the cultural heritage. All problems arise within a particular culture and are subject to socio-political and cultural forces. To cut students off from this heritage is to set them adrift in rough seas without any maps or compasses.

For example:

- What mathematical ideas would have been required by the builders of the pyramids or of Stonehenge? (Were there accountants keeping track of what workers were paid, fed on, and where they were housed? What knowledge did they need to work out where to place their stone blocks?)
- What mathematical ideas would have been required by people trying to build accurate clocks?
- What social forces led to or supported the invention of calculus?

In mathematics, connections are as important as specific results, for when working on a problem it is much more important that an idea or possibility come to mind than that the details of it are to hand; details can always be looked up. If you do not think to view a problem from different perspectives, you may not be able to resolve it satisfactorily, as is evident when attending any congress of mathematicians. One part of good connection making comes from paying attention to cognate mathematical domains; the other part comes from becoming more aware of the historical sources of your own problem and of related problems and techniques.

As an expert matures, they often find inspiration in the work of earlier generations and, through that scholarship, are able to make connections and to develop problems that have been overlooked. Thus, investment in the history of what you are teaching may easily pay off through becoming aware of the origins of your own research problems. For example, John Wallis, who wrote one of the first algebra books in English (the *Treatise of Algebra*, 1685) treated algebra as history. For him, to be at the forefront of research was to have taken on board what had happened in the field up to that point.

There are other advantages to drawing upon historical accounts, for the struggles that others have experienced are often experienced by students as well. Sometimes modern notation and modern ways of thinking can make what was very difficult for previous generations rather easy. For example, early attempts to solve quadratic equations by Babylonian, Egyptian, and Arab mathematicians were hampered by a need to keep coefficients positive. Thus they discuss three different kinds of quadratic equations where we are able to treat them all as being of one type (van der Waerden, 1983). Even now many students are tempted to write, 'Let $-x$ be a negative number', or to worry about x when it is supposed to be negative, wondering if they should indicate this by writing $-x$.

To take another example, Diophantos appears to have used only one variable in solving equations (Kline, 1972), whereas we are happy to use as many as required. However, some students of mathematics still find expressing generalities involving two or more parameters extremely challenging. The idea of a function emerged over many years, with many different interpretations (Wanner and Hairer, 1996), and still students are bewildered by the range of 'pathological' things that have to be admitted as functions. In such cases, students can marvel at the perspicacity of early mathematicians in carrying through lengthy and complex calculations, or struggling with cumbersome notation, while at the same time recognising aspects of their own struggles.

There are further advantages both to including historical remarks within a mathematics course, and to including a course in the history of mathematics within an undergraduate curriculum. Governments are increasingly calling for university courses to develop skills in communication (both written and spoken) so that graduates can function effectively in employment. It is vital for them to become accustomed to summarising ideas and to communicating these effectively, whether verbally in meetings or in written reports. This is true even for those who wish to become academics and teachers. Being able to extract the essential points of an issue is an important skill in any form of problem solving. Juxtaposing the development of such skills within mathematics and within a historical context provides a more balanced experience for students whose future success may depend as much on their presentation as on their logic.

How to Use History

A simplistic use of history is to include pictures of famous mathematicians associated with the topic in hand. Stories can be recounted about those mathematicians together with a brief indication of their other interests, for students often fail to appreciate that a mathematician whose name they meet in one topic actually worked in several other mathematical domains as well.

Tactic: Pictures and Stories

Tactic: Over time, develop a collection of portraits of mathematicians involved in the principal definitions and theorems of the courses that you teach, together with some remarks about their range of interests and any anecdotes that might interest students.

Many of these can be found on the MacTutor website (see the *Websites* section of the Bibliography). Remember that for most mathematicians prior to the 20th century we often have only one or two images, which is liable to give a static sense of the person rather than a sense of their development and maturation over time (for example the standard picture of Euler is as an old man). Alternatively, students can be sent to the MacTutor site, or to search the web themselves, and you can collect what they find for your own edification!

Tactic: Assign different mathematicians to different groups of students, and ask them to find out what they can about their mathematician. Have the results displayed on a notice board outside the lecture room or on a website.

You may be surprised at how much effort your students will put in when they know that their work will be displayed. Developing your own website, with not only links to standard sites but also extra content specifically constructed to enrich your particular course can provide students with the sense of participating in a developing enterprise.

Attention can also be put on topics rather than on mathematicians.

Tactic: Assign different theorems to different groups of students, and ask them to find out what they can about the origins of those theorems, and who else might reasonably have had their name attached to them. Similarly, ask students to trace the origins of a definition.

An historically-based task could replace one of the homework questions, or could be set as an extra assignment over a term.

In the history of mathematics, as in mathematics itself, what enthuses students most is some sort of a disturbance, a controversy or conjecture to be considered, modified and justified. This piques their interest, offers a focus of attention when they are in danger of becoming mired in details, and provides something to investigate or to find out about.

Tactic: Locating Earliest Known Uses of Symbols or Ideas

When was the ellipsis (...) first used? What did people mean by the terms triangle or function at different times? When did $\lceil 1.4 \rceil$ and $\lfloor 1.4 \rfloor$ come into use and why? When and for what purpose did subscripts enter mathematics? When did Pochhammer's notation

$$(a)_k = a(a+1)(a+2)\cdots(a+k-1)$$

come into use and in what context? The principal source I have used is Cajori (1993), but various websites are building up extensions and elaborations of this discussion (see the *Websites* section of the Bibliography).

Tactic: Invite students to trace the development of a notation which they are or will be encountering.

Tactic: Exploring Controversy

Not everything recorded on a website is necessarily accurate, and not always have mathematicians seen eye to eye. Students can be set the task of clarifying differing views found on different sites, as well as mathematical and philosophical controversies, offering their own opinion based on the evidence they can find. For example, they can investigate:

- the disagreement over whether mathematics is discovered, uncovered, or constructed;
- the disagreement over whether mathematics is best seen as an empirical process (Lakatos) or as a process of formalising informal ideas.
- the controversy over whether Babylonian tablets display evidence of awareness of generality;
- the controversy between Leibniz and Newton regarding which of them was the originator of calculus;
- the controversy between Fermat and Wallis regarding methods of discovering mathematical relationships involving sequences and series;
- the controversy between Kronecker and Cantor regarding the value of Cantor's ideas;
- the controversy between Kronecker and Dedekind regarding the real numbers;
- the reluctance of mathematicians to accept the formalisation of quotient groups (Nicholson, 1993);
- the difference between Fermat's and Huygens' approach to the problem of refraction (Tikhomirov, 1991, p19-26);

- the use of the 'principle of continuity' used, for example, by Poncelet to get his closure theorem (Bos, 1993, p131).

Tactic: Invite your students to describe the conflicting points of view succinctly but clearly and to offer their own opinion, by referring to historical sources, possibly starting from a website.

Fauvel and Gray (1996) is an excellent source of extracts, while Katz (1993 and 2000) provide exemplars of the use of history to inform the teaching and learning of mathematical topics. Several others are also mentioned in the *Historical Sources and Resources* section of the Bibliography.

The history of mathematics takes place within the broader history of the development of ideas. What was going on politically and culturally while mathematicians were developing their ideas? What influence might that have had?

Tactic: Making Links

Tactic: Set students the challenge of finding a sequence of links between two apparently disparate ideas in your course, or between a mathematical idea and some historical event or movement happening at the same time, by tracing historical connections as suggested by various websites or other historical accounts. (See also *Tactic: Linking Terms*, p84)

For example:

- in logic, trace the origins of Brouwer's mathematical constructivism;
- in group theory, trace the emergence of the notion of an invariance respected by a collection of symmetries (Möbius) through to Klein's characterisation of geometries and thence to algebraic topology, and uncover the different ways in which invariants have been studied;
- discover which mathematicians influenced a particular historical event such as the French revolution, the administration of Napoleon, or the construction of the hydrogen bomb, and how they and their fields of interest interacted with each other.

A more ambitious use of historical records can involve the use of original (or translated original) materials. As mathematicians, we depend on colleagues who make the history of mathematics their priority, but there are a growing number of sources for such material.

Tactic: Using Original Material

Tactic: Invite your students to read, make sense of, and summarise in modern notation a page or two taken from the writings of an historical figure. Invite them to read and rewrite in their own words a short article taken from a journal (such as the Classroom Notes section of the American Mathematical Monthly).

For example:

- Dedekind on the construction of the reals (Wolff, 1963, p138-169; Newman, 1950, p528–536);
- Wallis on general exponents (Smith, 1959, p217–218);
- Calkin and Wilf (2000), Recounting the Rationals.

See also the *Historical Sources and Resources* section of the Bibliography.

Reflection

There are many reasons for including some historical elements within a mathematics course, as well as for encouraging your students to study the history of mathematics in its own right. A few small steps, such as getting them to do some background research for you, will enable you to enrich your courses without making impossible demands on your preparation time. As you begin to find out more about the historical roots of the topics you teach, you may begin to find unexpected links to other topics, perhaps even those of research interest. You may also uncover problems that you wish to pursue and develop as a research interest.

Task: Reflection

Look up the background to one of the topics you are currently teaching, or are about to teach. What problems did that topic resolve? How did the major players get involved? Can you describe this to your students?

---

7 Issues and Concerns in Teaching Mathematics

'The authority of those who teach
is a hindrance to those who would learn.'
(St. Augustine)

Introduction

Studying the process of teaching and learning mathematics is very different from studying mathematics itself. In mathematics, problems can be posed (and often solved), theorised and formalised, and the solutions or solution methods can help to solve other problems, or to frame new ones. In mathematics education there are a number of endemic issues or tensions that cannot be resolved by any assertion or 'theorem'. There are no universal solutions, and it is wise to be wary of those who speak as if there are. Indeed, seeing tensions as problems, rather than sources of energy, is part of the problem.

One reason why there are no theorems in mathematics education is that the range of possible situations is enormous, and there is no way of gauging the sensitivity and awareness of the perceiver. Problems that some people see as similar are often seen as being different by people with different predispositions and perspectives, and apparently dissimilar problems are often described in similar terms. For example, what looks superficially similar, such as the behaviour of students from year to year, may actually be quite different when looked at from their point of view.

Another reason for an absence of theorems is that the objects of study are not ideas, as in mathematics, but people. People can choose to take the initiative or not. They have fallible and incomplete memories. They have values and desires, propensities and habits, interests and dislikes. They are often inconsistent and apparently irrational. Consequently, the logic of cause and effect, of delineating and controlling some factors while others are varied, just does not work on any significant scale.

Despite all this, people still, on the whole, want to make sense of the world in which they find themselves. If they do not appear to want to make sense of mathematics, it may be because they have already made sense of the overall situation and concluded that it is not important to them. However, the basic hypothesis of this book, and indeed probably of every teacher of mathematics, is that almost all students *can* be attracted to making sense of mathematics, especially if their sense-making powers are called upon and made use of. There is nothing so boring and deadening as having someone else do all the work for you. Equally, however, it can be depressing to be overwhelmed by too much work. Striking a balance is what making use of tensions is about; not neutralising or minimising them, but rather moving about on a spectrum of possibilities so as to get energies flowing. If tensions are perceived negatively, as problems to be removed or circumvented, then they will probably continue to produce frustration and anguish. If they are seen as sources of stimulation, on the other hand, then they can perform a positive role. For example, if, through their continued presence, you are constantly or frequently reminded of the existence of one or more tensions, then this awareness can keep you awake and able to making fresh choices, rather than falling into a routine. This means that you are likely to remain observant of and sensitive to your students.

There is, however, a sense in which there *are* theorems in mathematics education, if you return to the etymological roots of the word *theorem.* It comes from a Greek word *theorein* meaning 'look at', or perhaps 'a way of seeing'. Thus a mathematical theorem can be thought of as a specification of what to 'see' and how, and its proof as an attempt to get others to see similarly. Now, a great deal of mathematics education can be seen as being about ways of perceiving teaching and learning situations, and thus as being about 'theorems'. Theories in education provide language in which to describe rather than principles from which to deduce how to teach. The analogues of proofs are attempts to get others to see issues and actions similarly. This book has focused on actions, but in the process of trying out some of them for yourself I hope that you have found yourself making finer distinctions, becoming sensitised to details which you previously may not have noticed, appreciating alternative perspectives on teaching and learning, and perhaps even being stimulated into becoming more articulate about what you do, and why, when you teach your students.

This chapter revisits many of the ideas of the previous chapters. No new tactics are introduced; rather, it attempts to distinguish some fundamental ideas that underpin and inform the many suggested tactics in the other chapters. The chapter begins with a discussion some endemic tensions, clustered under three main headings but closely interwoven.

- *Student and tutor agenda and expectation* This section encompasses issues of time (coverage and pace), personal viewpoint (serialist or holist), and challenge (insufficient or excessive) that will be developed later into an implicit contract.
- *Doing and construing, knowing and understanding* This section encompasses an important distinction between doing and construing, and how this affects what students believe they have to do, know, or understand.
- *Being subtle and being explicit* This section encompasses issues concerning excessive and inadequate intervention by tutors, and effective interactions between tutors and students.

There then follows a discussion of several issues of that are topics of ongoing debate.

- What is exemplary about examples?
- What is the place of definitions in teaching mathematics?
- Is mathematical discussion possible?

In order to make use of the energies stored in these tensions it is useful to call upon two major resources: the powers that students bring to their learning, and your awareness of mathematical themes and processes. Some of these are elaborated in the third section, while the final section describes three useful frameworks for structuring your preparation.

Tensions

Tension: Agenda and Expectations

Students arrive in class with their own agendas. For example, they may want to minimise their investment, or understand the subject matter more deeply. They may even have chosen the class for some other reason that has nothing to do with mathematics. Tutors also have their own agendas. It may, like their students, be to minimise their investment of time and energy so as to be free to do research or to teach other courses. On the other hand, it may be to view some of the ideas afresh and to take pleasure in appreciating and ameliorating some of the difficulties that their students encounter. It may also be to force themselves to learn or review the subject matter.

A tutor who comes to their class wanting to be very helpful may be frustrated by the responses of students who 'just want to pass', while a student who really wants to get to grips with the ideas may be frustrated by a tutor who is minimising their own investment of energy and time, or treating students as if they want only to pass. Consequently, there is more to achieving harmony with your students than simply announcing your own agenda. It may require an adjustment on both sides, and this may take time as it is quite hard to do explicitly. Instead, you have to induce your students to want to put in time on your course without demanding so much that they have too little time for their other courses. They, in turn, have to attract you to want to help them, through the way that they participate in class and through their homework assignments. However, they may not be explicitly aware of these possibilities. An agenda sets up expectations on both sides, and it is those expectations that can be worked on explicitly.

If you choose not to meet those expectations, you need courage and personal confidence that your students will come round to your way of thinking. Rather than forcing one particular way of working on your students, you can use a mixture of styles, some that you think are effective, and others that they seem to expect. You can then hope to wean them onto your preferred styles where they prove to be effective. However, this will not work if you have decided in advance what will be most effective. If, instead, you see in each group of students an opportunity to 'negotiate' a way of working suitable to them and to you, then both you and they are likely to be happier with the results.

For example, John Berry (1999) reported that in a third-year modelling class he adopted a structure in which groups of students carried out projects and presented their findings to the class, while he observed, supported, and questioned them. This has some similarities with the R. L. Moore method (p102), but it is less competitive and less tutor intensive. At first there was considerable reaction against the 'new format', but, by the end of the year, most of the students recognised that they understood the material much more thoroughly than in their other lecture-based courses, and they even found that they needed less preparation for the (standard) examination.

Students expect their lecturers and tutors to tell them what to do, and that if they do it then they will pass, get their degree, and be admitted to their chosen profession. You can reasonably expect your students to do the work you set, to gain facility in the techniques, and to take an active stance in making sense of the material you present. Some degree of explicitness at the beginning of term about these expectations can be useful for your students as well as clearing the air between you and them. For example, some lecturers distribute a sheet in or before the first session, laying out such expectations (Siu Man Keung, reported in Mason, 2001).

Three instances of the gap between student and tutor expectation arise as familiar tensions:

- time (coverage and pace);
- personal propensity for viewpoint (serialist or holist);
- challenge (insufficient or excessive).

I will return to the implications of student and tutor expectation after elaborating on these three tensions.

Issue: Time – Coverage and Pace

'My job is to lecture and yours is to listen.
Let me know if you finish first!'
(Paul Halmos at the beginning of a series of lectures in 1972)

'There is so much to cover, so little time in which to do it. If I do half the things you suggest, I will never finish the course.'

Shortage of time induces a tension between coverage and pace: you must cover the syllabus, but you want to do it at a pace that enables your students to learn what they have to know for the examination, and hopefully also to appreciate and enjoy the ideas involved.

This may lead you to ask more questions. What constitutes 'the course'? Are your notes the content of the course, or do your exercises and assessment tasks form the content? Does mathematical thinking play any part in what constitutes the course as perceived by your students? Most students start by seeing the lectures and texts as the course, but when exams loom they switch to seeing the course as being defined by the assessment.

Is it only after you have said something that your students understand it, study it, *encounter* it? If you say something, does it follow that your students are in a position to hear, integrate, or make sense of what you have said? Is it more important that *you* finish the course, or that your *students* finish it with some degree of comprehension?

You can 'cover' a course without actually saying everything yourself, but only if you have supported your students in learning how to learn mathematics. Then, as they become better at employing their own powers to make sense and to think mathematically, you can concentrate on providing the most helpful generic examples, exposing wrinkles and subtleties, and both motivating and proving theorems.

If you try to cover the whole course at a uniform pace, then your students will become used to that pace and will adapt to it, doing no more than is required. If, instead, you vary your pace, slowing down to work on ideas you know to be difficult then speeding up and explicitly calling upon your students to work on more routine matters themselves, then you are more likely to find that both you and your students complete the course satisfactorily. The changes in energy will be more likely to keep your students more awake and enthusiastic about their studies.

Issue: View – Serialist or Holist

Some people like to have, even need, an overview before they begin; before they set out on a journey, they have to know where they are going on some map. Others are content, or even prefer, to be led along step by step. Both propensities can lead to difficulties. It may not always be possible to explain where one is going if the territory is strange and the technical language unfamiliar, or if the very act of being explicit will render it unattainable (for example, telling people in advance that you are going to give them an impossible task because they need to learn how to prove that something is impossible, as well as to prove it is possible).

Expecting your students to trust you and to follow whatever you do is probably unreasonable, especially if (or when) they begin to lose their way, to lose sight of the woods because of all the trees. However, sign-posting need not always be done at the beginning: it can be refreshed as and when your students give indications that they need it. Sign-posting can be particularly effective when it is local, as in, 'In this proof we will see the use of ...', or, 'This technique provides the foundation for this topic.' Global sign-posting provided at the beginning of a course can all too easily be forgotten when the going gets tough later. Several of the tactics suggested in the earlier chapters are designed to provide your students with relevant direction when they need it. Pausing and being explicit about unexpected connections can be very helpful. For example, you could compare and contrast the intermediate value theorem in real analysis with Cauchy's theorem in complex analysis, or even ask your students to make the comparison, or you could compare and contrast different approaches to approximating functions (via Fourier series, Chebychev polynomials, hypergeometric series, and so on).

There is an inevitable tension between expecting your students to trust you, simply following wherever you lead, and trying to tell them what the enterprise is about, and where things are headed. Instead of resolving always to provide advance organisers and summaries, or always to expect your students to follow your trail of definitions, theorems and proofs, you can use your awareness of this tension to pause at various places en route to discuss an overview of the landscape or to make some connections before diving down again into detailed logical development.

Issue: Challenge – Insufficient or Excessive

Halmos (1985, p271) said that, 'Challenge is the best teaching tool there is, for arithmetic as well as for functional analysis, for high-school algebra as well as for graduate-school topology.' However, if the challenge is too great, students may be put off, while, if the challenge is too slight, they may never become active, never transform the words and symbols into an expression of what they think or can do. They may be lulled into a false sense of security, or simply become bored. There is always a tension in attempting to strike a level that provides a challenge without being too extreme, but it is better to try, and to make adjustments, than to try to do the learning for your students by making it all too simple, clear, and straightforward.

Varying the degree of challenge is a far more fruitful strategy than making everything easy for your students, or making it all very difficult. They will soon work out which problems they can make progress on themselves, and which they will require extra help in order to complete. As long as you are clear about what is required to pass, to do moderately well, or to do very well, your students will respond accordingly. Most importantly, it is wise not to assume that you are the best judge of what is challenging and what is not: one of the major forces in student non-achievement is teachers being all too certain about what they think their students are capable of. Despite our competitive culture, relatively few students respond to being told they will not succeed by striving harder. Put another way, motivation is not something that you can do for your students, but something they have to do for themselves. Motivation is a state; to be highly motivated is to be eager to take the initiative, to fashion and mould things the way you want them to be, and to try to get to grips with ideas and behaviour. What you *can* do for your students is to provide the kind of context, environment, and way of working that encourages them to motivate themselves. This idea lies behind the tactics concerned with establishing and maintaining a conjecturing atmosphere.

Issue: Implications of Student and Tutor Expectations

Because of these expectations, there is always a contract between teacher and student, which may be more or less explicit, and gives rise to an endemic tension, mentioned in Chapter 4, articulated by Guy Brousseau (1984, 1997).

- The more clearly and specifically the teacher demonstrates to the students the behaviour being sought, the easier it is for the students to display that behaviour without generating it for themselves from understanding the material.

For example, the more nearly assessment questions match examples done for homework or in class, the easier it is for students to find a matching example and change the numbers. The more precisely a lecturer indicates what will be examined, the easier it is for their students to prepare for that examination by using only short-term memory. If, on the other hand, the assessment is unlike the homework or the class examples, if the examination is a complete surprise to students, then their work is depressing to mark and they may reasonably complain that their lecturer has been unfair.

Finding a balance between these is difficult, for this tension seduces the lecturer into making things easier for their students, with the effect that they can appear to understand but may be working only from trained behaviour. Similarly, as discussed in Chapter 4, structuring tasks too elaborately can remove the need for students to do anything more than complete each sub-task in turn (Nardi, 2000b).

Another way to describe this fundamental tension is that training in behaviour is fine (it is useful for students to gain facility in using standard techniques on standard problems), but tends to be inflexible. As long as the assessment questions are identical to training exercises, your students will survive, but if you want them to appreciate mathematics, to think mathematically, and to cope with unusual or varied problems, then their awareness needs to be educated as well. They need to be drawn into working on mathematics, not just rehearsing routine exercises.

The more you give in to your students' desire for worked examples, the more likely you are to provide them with only enough to master the superficial techniques, but not enough for them to integrate them into their awareness. If you stress theory over technique, your students may memorise proofs of theorems line-by-line, but if you stress problem technique over theory, they may master question types as templates with little appreciation for why or how they works or when to employ them. Anything unusual will throw them off completely.

Guy Brousseau described this tension in terms of an implicit contract (*contrat didactique*) between a group of students and their teacher, developed in their first few sessions based on past experience. Students agree to do what they are asked, on the understanding that this will get them a pass (or a first), and constitute learning. Such a contract is an endemic and ubiquitous feature of teaching.

It may be helpful to be explicit with your students about some aspects of the contract. For example, being explicit about how you are going to structure sessions may not only help some students to free up time, but may also intrigue them and attract at least some into participating more fully. For example, faced with a class that was due to meet five mornings a week (at 7:45 am), I told them that we would finish a chapter a week; that on Mondays I would work with them on the ideas, that by Thursday we would finish the chapter, and that on Friday we would do revision questions. I saw the essence of the contract as being that I would start each week working the way I thought was valuable, but would guarantee to work the way they (currently) thought was most valuable in order to finish each chapter. I soon found that I was working 'my way' until part way through Wednesday, when I would cut and run, giving a brief exposition of the remaining ideas (if any). Far from skipping Monday's class, more and more students seemed to appreciate this way of working on the ideas.

There are, however, significant difficulties with being explicit, which I will take up shortly. Meanwhile, there is much more to say about the issue of what students actually do and what sense they make of it all.

Doing, Construing and Wanting

Modern psychology recognises three fundamental aspects of the psyche, already present in ancient Indian writings:

- cognitive (intellect, thinking, awareness);
- affective (emotion, motivation);
- enactive (behaviour).

These three aspects show up in many guises, and it is through drawing upon and integrating their strengths that effective teaching and learning take place. One manifestation of this emerges from the *contrat didactique*, which distinguishes between students 'doing tasks' and construing or making sense of the mathematics involved. The energy for these comes from each student's desire and motivation.

Issue: Doing is Not the Same as Construing

Just doing exercises does not necessarily equip students to do similar exercises in the future. The usefulness of exercises depends on whether your students make use of the exercises to reconstruct general techniques and theorems, or whether they simply try to complete the exercises. Making sense of a technique, theorem, or topic requires more than passively 'suffering' the tasks as set; it requires an active and enquiring stance, a desire to make sense of it all.

Naturally enough, students taking mathematics as a service subject often want to reserve their energies for their main interests. They often see mathematics as a diversion and a burden. You can accept their arguments and collude with them to reduce their effort by giving them lots of exercises to do, but you can also discuss with them how to work on mathematics, because it is likely to influence how they work on their own discipline.

Students who simply work *through* tasks one by one, who are satisfied that they have 'done the task set' even if they could not do it again in the future are not in nearly as good a position as those who work *on* their assignments. Your students will gain much more by working *on* problems, trying to see how they illustrate or extend a theorem or technique, trying to see what constitutes a question 'of this type', and possibly even trying to construct other examples like it. If they can satisfy themselves that they could do other questions like it in the future, they are much more likely to gain proficiency and also make sense of what they are doing. If you demonstrate mathematical thinking both with and in front of your students by being explicit about what constitutes working *on* exercises, you can assist them to tackle the learning of mathematics like a mathematician. This is the aim of many of the tactics presented in earlier chapters.

We have all met students who seem to understand and yet are not very good at doing problems, and others who seem to have facility, at least with standard problems, but do not appear to understand what they are doing or why (and even if they appear to understand, they are likely to tell you that they do not). This applies particularly to topics that your students have studied in previous years!

This produces a difficult tension: if you had to choose, would you rather your students became good at the techniques or understood the concepts? Of course, a balance between the two is probably the most sensible, but what would you do if you had to choose?

Most lecturers would probably like to teach their students the mathematical concepts, to get them to understand the ideas. The theory would be that students could then reconstruct anything they forgot, whereas students with high technical proficiency might not recognise an opportunity to use a technique in an unfamiliar setting, and would be unlikely to be able to adjust the technique to meet fresh circumstances. This ability is vital, especially for mathematical techniques that are to be employed in other disciplines.

Understanding clearly requires some degree of technical proficiency, however, at least in the techniques needed to reconstruct theorems and techniques! To test understanding, you are likely to set your students problems to solve, which also require technique, and/or theorems to prove (whether reconstructed or from memory). So some technical proficiency is desirable, though it is a common conjecture that this will somehow 'come with understanding'.

Whatever teaching method you use, it is important to set tests that are fair to your students. When students are given novel and unfamiliar questions on a test, they tend to do badly. If they have not had the experience of being challenged in this way throughout the course, they are very unlikely to succeed under pressure.

Some lecturers enjoy their own high degree of competence and facility with technique and expect something similar from their students. They may have gained facility themselves with little effort, and consequently cannot work out why their students do not put in sufficient effort to gain mastery. They may not realise the extent to which their understanding of the underlying ideas enabled them to recall and reconstruct on the fly, and they may not appreciate the struggles of those who find mathematics difficult. Catching yourself struggling at something and exploring the possible parallels with your students' experience is the most effective way of developing sensitivity to their difficulties.

Some lecturers recognise that students, especially those on service courses, really only want to pass the course with a minimum of investment in order to 'get what they need'. It is tempting to collude with this view and try to pare down the content to a collection of basic techniques. Students can then be given large numbers of practice problems, perhaps even supplied via a computer system, the underlying theory being that 'practice makes perfect'. However, there are difficulties with this view as well. Students who confine themselves to memorising techniques often find that their memories do not last long, and they are notoriously bad at recognising the relevance of a technique in a novel context. Students who simply work their way through exercises may or may not be learning anything.

Experts in areas that use mathematics sometimes insist that 'students need to learn the mathematics that I know, in order to be expert like me', yet it is rare to find two experts in the same domain who agree on what mathematics is essential. The reason must be that each person is shaped by the particular area of expertise they develop. What students really need is the confidence to be flexible, to be able to search out what they need, make sense of it, and apply it as necessary.

There is an endemic tension between technique and concepts, between understanding and facility, and hence between a pedagogy based on ideas and a pedagogy based on practice.

For example, suppose you have been working on Riemann integrals with students coming from physics or engineering, and feel they are ready for applications. You get them to work on finding areas (choosing horizontal or vertical strips, checking endpoints of intervals on which one function is above another, checking continuity, and passing to the limit). You indicate the general technique of decomposing the integral into intervals between the crossing points of the functions and then using the integrals for these straight away. You then move to finding moments and using these to locate centres of gravity. Again you work with them (you demonstrate, asking them questions as you go so that they feel they are participating) on setting up strips, writing down the moments for each strip, checking continuity and intervals, and then passing to the integral.

What might they have learned? Some may have been actively thinking about the problems and recognising elements of the theory of Riemann integrals. They may have abstracted the elements of the method. Some students may have been mystified by the use of strips and the calculation of a moment, so that passing to the integral is out of reach for them; these same students may earlier have focused on the methods of integration but avoided the theory. Some students may have struggled to keep up with the exposition, having adopted a passive 'copy it down and work it out later' approach.

Simply presenting a careful exposition, working through examples, and developing ideas with applications do not in themselves guarantee successful learning. An important element of learning lies in the hands of your students. The teacher's role is to encourage their students to use their own powers to work on the mathematical ideas.

Ideally, technique and understanding will develop together, for improvement in technique frees attention for understanding, and understanding enables techniques to be reconstructed. However, students often seem adept at separating them, at gaining some degree of facility without really understanding what they are doing. Of course, full understanding often only comes when someone is called upon to teach someone else; this is when they are pressed to reconstruct the ideas and theorems fully, for themselves as much as for their pupil. This is why most of the tactics proposed in this guide are variations on stimulating students to reconstruct and express things for themselves or to their colleagues, either in sessions, with each other, or by themselves.

Understanding and technique are not easy to test separately. When one lecturer looks at the assignments set by another, they usually see the questions as being technique oriented, even when the person setting them feels they challenge conceptual understanding. You can aim to set conceptually easy questions that are very demanding on technique (for example, 'Show that, for x and y positive, $x^x + y^y \geq x^y + y^x$.') or technically undemanding questions that require an appreciation of the concepts (for example, 'Show that if A and B have exactly the same complete set of eigenvectors, then AB and BA also have those same eigenvectors; find the eigenvalues of AB and BA in terms of the eigenvalues of A and B.'). However, it may be the case that, to a student, the two are intertwined; that if a question employs a concept that is still fragmented, still in the early stages of being enriched, exemplified, and connected to associated techniques and contexts, the associated technique can seem difficult, and vice versa.

Issue: Knowing and Understanding

What does it mean to understand mathematics, or a particular topic in mathematics? What does it mean to know how to carry out various techniques, or when to use a particular technique? In contemplating these questions I find myself asking, 'In what sense do I understand numbers? The more I think about them, the less I seem to understand them.' However, this is not what I mean when I say that a student does or does not understand numbers.

What does 'understanding' mean to students and tutors? Almost certainly students and tutors have different interpretations. This tension has been studied in some depth, with as many different approaches emerging as there are authors writing about it. (See for example Sierpinska, 1994, or Pirie and Kieren, 1994). Richard Skemp (1976) developed a much-used distinction between *relational* understanding (appreciating inner structure and connections with other topics, uses, and so on) and *instrumental* understanding (facility with technique with little idea of why it works). He went on to develop theories about how and why students end up with instrumental understanding when their teacher expects relational understanding. Most definitions of understanding in the literature are cast pragmatically in terms of what someone is able to do (or what they have done in the past). Only more rarely do you come across the view that understanding is a complex personal state of self-confidence, where you know from experience that you can trust ideas to come to mind when needed, and, when they do not, you feel able to admit it.

Understanding is tested and developed by doing questions, working through examples, having questions answered, working through proofs and techniques on examples you construct for yourself, solving non-routine problems, explaining to a fellow student, and so on. Consequently, a system that encourages students to interact with the material, each other, and their tutors in order to experience a range of these is most likely to support them in developing understanding.

It is one thing to set up a system intended to support students. It is quite another to arrange that they get the maximum benefit from that system. Students need to appreciate what different kinds of sessions or different support agencies can offer them. They can also be all too easily overwhelmed by excessive workload.

Being Subtle and Being Explicit

Not everything can usefully be spoken about explicitly in the context of teaching and learning. For example, in Brousseau's analysis, mentioned earlier (p166), it is essential that no matter how explicit one tries to be with students about ways of working and the ideas being worked upon, the *contrat didactique* must remain implicit, for otherwise it leads to failure. As mentioned earlier (p130), Tahta (1980) put forward the case that every task or activity has an explicit or outer aspect (what people are asked to do, what they do in response) and an implicit or inner aspect (what people are likely to encounter in the way of details of the topic, mathematical processes, heuristics, and their own personal propensities). If the inner aspect becomes outer (is made explicit) then the whole nature and purpose of the task is lost, just as when you are working on a problem and someone drops a clue or tells you an answer and you feel deflated and uninterested in continuing. Students may then be able to circumvent the work they need to do. They may think they have done what is required by going through the motions ('If I get answers to the questions the teacher sets, then somehow I will learn.'), but in fact they may not make important connections, form important images, or to come across the wrinkles, and perhaps even fail to re-construct core ideas to displace their previous conceptions. Brousseau points out that what is most important about the implicit contract is that it can be ruptured, for it is when there is a disturbance to established habits that interesting things start to happen (Brousseau, 1997, Chapter I, section 3.2).

If everything is made explicit then students are supported in becoming dependent on the teacher to do things for them, but if too much is left implicit many students may never discover what is being made available to them. This applies at every level: to content, to mathematical processes, and to heuristics. The issue is not so much whether to be explicit, but rather when and to what extent students can benefit from it.

The problem with being subtle and implicit is that many of your students may simply never become aware of the undercurrent, of the essence of what is being offered. On the other hand, if you are too explicit you may make it difficult for them to come to their own understanding of a topic or to internalise it in their ways of thinking, and so be stuck with a superficial memorised version of what they have been told. For example, if a tutor constantly varies the type of question they ask, their students are not likely to notice that the questions are all variations on one or two themes. They are unlikely to become aware of the range of types of questions, and even a shift of their attention to the types of questions being asked ('What question am I going to ask you?') is unlikely to produce much awareness. However, if they tutor always asks the same type of question over and over, their students may become inured to those questions, and become bored and frustrated, even dependent upon their tutor. The students may also form the impression that the questions demonstrate the totality of what mathematical thinking is like.

Some people find it natural to be explicit about their thinking processes, about what they attend to as they do mathematics. Others fear that, if they are too explicit, their students will not internalise what is being pointed out, but rather confuse it with the ordinary content. This is an endemic tension to which there is no answer. Being aware of it can keep you sharp, keep you thinking about what it is that your students are attending to and what they are missing. It can also enable you to choose to be explicit sometimes and more subtle at others (see *Tactic: Directed – Prompted – Spontaneous*, p91). As with any endemic tension, the way to use it positively is to resist settling on any one resolution, and to use any reminder of the tension to alert yourself to possibilities for exploring other approaches. This is what lies behind the tactic of gradually reducing explicit interventions of any one type, being explicit and directive to begin with, but then increasingly indirect and subtle until your students are spontaneously using them for themselves.

A useful time to be explicit is when you are doing a worked example or proving a theorem in front of your students, as suggested in a number of the tactics in earlier chapters. Try to be aware of your 'inner monologue' and thought processes, and try exposing these to your students.

Issues

There are a considerable number of ongoing issues in teaching mathematics that, like the tensions mentioned above, are unlikely ever to be resolved. It is nonetheless valuable to try to articulate them, so as to stimulate your colleagues to describe their practices to you, from which you can pick up some useful possibilities for your own teaching. I have chosen four issues, two of particular interest to me (the exemplariness of examples and the place of definitions), and two more general issues (the place of discussion in learning mathematics, and the issue of motivation). This is followed by the suggestion that students bring the requisite resources to their studies, but that it is the tutor's job to make sure that they actually make use of them.

Issue: What is Exemplary About an Example?

'Affirmative or negative, it's examples, examples, examples that, for me, all mathematics is based on, and I always look for them. I look for them first, when I begin to study, I keep looking for them, and I cherish them all.'

(Halmos, 1985, p64)

When students are offered an example of a vector space, group, differential equation, or Fourier transform, they often do not recognise what makes the particular object exemplary, much less see what it exemplifies. When they try to approximate a function in a space of polynomials, it is often very hard for them to 'see' the problem in terms of a geometrical picture of a vector space and the calculation of a projection. The problem is that if you do not yet appreciate or comprehend a generality, then it can be hard to see how something is an example of it, yet the example is supposed to help you to understand the general! Which features of the example are generic and which are particular to the special case? Matters are complicated by the fact that students often attend to unexpected features of the 'example', not the features you believe you are stressing!

For example, when an expert considers the eigenvalues of a three by three matrix, three plays no significant role in their thinking, as they know that the same processes apply to a square matrix of any size. They see through the three to the general. For a novice, the three may be invisible, playing no role in their computations but also not giving them any impression that it could be four or five or a hundred. On the other hand, the three may be overly significant for a novice who assumes that the calculations only apply to three by three matrices.

Seeing through the particular to the general requires a kind of letting go, a kind of blurring of the particular, so that it becomes a placeholder. However, this requires an awareness of the sorts of things that the symbol could be replaced by. In the case of the matrix example, an expert has a sense not only of 'any square matrix', but also that the three by three case is large enough to display most of the possible relevant behaviours of larger cases, while the two by two case is too small.

Hand in hand with seeing the general through the particular goes seeing the particular in the general. An expert can look at a general statement about eigenvalues of matrices and immediately imagine specific cases, and can construct particular examples illustrating different features of it.

Thus, what makes an example exemplary is an accompanying awareness of what is particular to the example and what is generic, what can be varied and in what way, and what must remain invariant. It follows that in order to help your students to experience an example as exemplary, it might help to be explicit about what features could change (and in what ways), and what is invariant among all examples. In this way, exemplification can be seen as an example of the mathematical theme *Invariance Amid Change* (p192).

Appreciating the general and how an example exemplifies it is part of the process of abstraction. The next step is to list the salient and relevant properties of one or more examples, and then declare these to be the sole properties required of an object in order for it to be an example of such a structure. This switch in thinking from delineating properties to declaring a list of properties as characteristic or defining is, for many students, a difficult one. This shift that is, of course, the essence of abstraction and axiomatisation, but students rarely get to experience it. Rather, they are often just presented with the results of such a shift made by someone else.

Students arrive in class with the power to abstract, specialise, and generalise, as will be argued shortly. Your task as tutor is to activate those powers. The tactics presented in the earlier chapters of this book provide a number of different actions you can take to support your students in making these shifts:

- show your students how to specialise in order to make sense of a generality by doing it yourself in front of them;
- pay attention to the features of an example that make it exemplary, and make sure that your voice tones and written work emphasise those features;
- when offering a worked example, be explicit about the choices that you are making at each stage, so that your students have a process to engage in, rather than simply a format to follow.

Becoming aware of what you are attending to, of what you are stressing, and then making sure that you communicate that to your students can make a big difference to their appreciation of mathematics.

See *Tactics: Boundary Examples*, p14, 136; *Student Generated Exercises*, p16.

Issue: The Place of Definitions, Theorems and Examples In Mathematical Exposition

Once the mathematical ideas have been worked out, an exposition of them usually begins with definitions, and proceeds, via lemmas and examples, to theorems and their proofs. However, definitions are often only made precise when there are some theorems to prove. Indeed, it may take several attempts at a proof before the conditions necessary for it to succeed are arrived at, and this often influences the definitions used (Lakatos, 1976). Definitions (and associated theorems) only survive if they proves fruitful for further study. Once the topic or domain is developed, the temptation is to 'haul up the ladder' and proceed as if the definitions were entirely natural and the theorems a natural consequence.

Bertrand Russell suggested that definitions are merely the renaming of a complex expression by a succinct label so as to make it more manipulable. While this is part of the story, before your students can manipulate the definition label with facility, it must actually mean something to them! Thus your students may need assistance in learning how to replace a technical term by an expanded version that inspires them with confidence (see *Framework: Manipulating – Getting-a-Sense-of – Articulating*, p187).

It is actually quite hard to take in a new definition without some sense of what it is trying to achieve, and very difficult to appreciate the ramifications of its details. Some lecturers prefer a polished mathematical exposition beginning with definitions, perhaps motivated by examples, while some recommend starting with the major theorems and uncovering the details required of the definitions through the judicious use of examples. Some and even go so far as trying to construct proofs with inadequate definitions in order to reveal to their students where the proof requires something more precise to make it work.

The tactic *Intensive and Extensive Definitions* (p25) suggested that it can be useful to work explicitly on the formalisation of intuition when starting with an intensive definition. The reverse process also makes sense, especially since in the future (and in texts) your students will encounter formal definitions. By working explicitly on how to make sense of an intensive definition or understanding, or an intuition about a technical term, students learn how to learn.

Working explicitly means looking for examples that fit, and examples that in some sense only just fail (see *Tactic: Boundary Examples*, p14, 136). It also means trying to form some sort of an image, picture, or metaphor to augment a succinct symbolic expression. After all, symbols are only a formal expression of some image, picture, or understanding. Finally, it means trying to explain your understanding to someone else, trying to bring it to in words without merely repeating the symbol sequence.

See Leikin and Winicki-Landman (2000) for a study of prospective teachers encountering different types of definitions and dealing with them in different ways.

Issue: Discussing Mathematics

Can mathematics actually be 'discussed'? Can students discuss mathematics fruitfully? At first thought, mathematics does not seem amenable to discussion in the way that philosophy or other arts subjects might. However, discussion includes more than exchanging opinions or arguing some position. It can also mean exchanging and refining interpretations, images, and examples in a collaborative attempt to clarify ideas and correct mistaken images or conclusions.

Most mathematicians, have at one time or another, walked into a colleague's office to ask for some help, described the (mathematical) problem to them, and then left (perhaps even saying thank you), without the colleague having to say anything. Articulating a difficulty to someone else is often the best way to uncover and even sort it out. However, articulating is much easier if there is an audience. It seems reasonable that students could play this role, and more, for each other.

What if students pick up wrong ideas from each other? If you believe that students learn what others tell them, then merely by their presence in lectures they ought to learn. However, it is not this simple for most students, so there must be more to learning from being told than meets the eye. Expressing your sense of a topic or problem, and listening to others struggling as well, is an integral part of coming to understand and appreciate mathematical ideas, because it is in trying to articulate your sense of relationships and connections, meanings and significances, that things fall into place. Thus, even though what students say may be incomplete or even partly wrong, the effort to express their thinking is an important contribution in 'getting it right'.

Effective discussion requires a 'conjecturing atmosphere', in which everything said is said with the intention of modifying it with the help of others, and those who are most confident give way to those who are struggling to express themselves. If you know you know the answer then, unless pressed, there is no need to tell others. On the other hand, if you are unsure or confused, there is every reason to get your thoughts out in the open so that they can be examined and modified in the light of other people's suggestions. Conjectures which tumble around inside your head are much more difficult to challenge or clarify than conjectures that have been expressed. (See *Tactic: Getting Students Used to the Idea of Proof*, p53)

Generating discussion may seem daunting at first, but there are ways to help get discussion started, such as the *Tactic: Generating Discussion* (p74).

Issue: Dependent and Independent Learners

Students depend on their tutor to structure the exposition, to provide suitable tasks, and to provide an exemplar of expert mathematical thinking, among other things. However, if students grow to depend on their tutor for everything, then they may not develop the independence they require to use what they learn outside of the classroom, or to learn more when they need it in some other context. In order to learn effectively, students need to learn to do for themselves what their tutor initially does for them. The whole thrust of the tactics proposed in this and the other chapters is to stimulate students to take initiative, to become less dependent, and to learn from their tutor and their tutor's way of working not just the mathematical content, but how to work at and use mathematics.

It might seem, then, that being explicit to students about what they should and should not do ought to be beneficial. It may indeed make them happy and allow them to hand in their homework with ease, but it may not serve them well in the long run. If a tutor establishes a consistent pattern of behaviour, so that questions are always couched in the same form, topics are always introduced in the same way (from the general to the particular or vice versa, from an application to the theory or vice versa) and sessions always have the same format, then students will become comfortable, but are likely also to become dependent. On the other hand, if the tutor is always springing surprises, if the questions change format constantly, and if there is no discernible rhyme or reason to what they are doing, then their students are likely to find it difficult to develop the necessary confidence.

Szydlik (2000) found that students who based their trust in external authority (their teacher or textbook) were very likely also to have an instrumental approach to studying calculus, as a subject which they had to memorize. They found it difficult or impossible to explain why limits work the way they do, seeing them as either unbounded or as unreachable, and struggled with problems involving them. On the other hand, students who based their trust in internal authority (their own convictions) saw mathematics as supposed to make sense, and were frustrated in class when they were not told why things worked. They, by contrast, were fairly successful in solving limit problems.

Dependence and independence are intimately tied up with the tension between being implicit and being explicit. If everything is always explicit then students are supported in becoming dependent on their teacher to make things explicit, but if too much is left implicit then many students may never encounter what is being made available to them. This applies at every level: to content, to mathematical processes, and to heuristics. The issue is not so much whether to be explicit, but rather when and to what extent the students can benefit.

The issue is not one of mere dependence, but how that dependence manifests itself. Certainly students will continue to depend on the tutor for access to further mathematical ideas and techniques. However, if they also become dependent on them to structure their work by providing lots of sub-tasks with every problem, or if they depend on the examination having a well defined format (rehearse definition, prove theorem, apply theorem), then their tutor is actually doing them a disservice, even though with the best of intentions. This is yet another version of the endemic *contrat didactique*. A useful device for exploiting this tension is the slogan 'try to do for your students only what they cannot yet do for themselves'. Sometimes this means being more challenging; sometimes it means waiting despite a strong desire to move things along more efficiently by making a suggestion or taking control. Supporting the slogan is the notion of *scaffolding and fading*, derived from ideas of the Russian psychologist Lev Vygotsky.

Scaffolding has two related meanings. It can mean holding up something (an idea, marker, position, computation, image, or plan) for your students in order to enable them to put their full attention where it is needed and to succeed at what they are doing. It can also mean a structure or intervention that enables your students to function, but which, if repeated too often, is likely to induce them to become dependent on you. This is what *fading* is for: the tutor at first directs their students' attention explicitly, but over time uses increasingly indirect and subtle prompts until they eventually act more or less spontaneously for themselves. When this happens, the scaffolding and fading is complete, and their students have internalised what was previously external and explicit. Vygotsky's suggestion was that students have to experience the thinking of other, more expert, thinkers in order to internalise something similar for themselves.

For example, at first a tutor might specifically and directly ask their students to make up a problem like the homework questions. Gradually this might shift to something indirect such as, 'There will be some marks for extending the problem set.' and become, 'Don't forget what 'we' do with homework problems.'. To take another example, a tutor might warn their students explicitly about integrating past an asymptote or watching out for the function crossing the axis when calculating areas, and then offer less and less direct reminders. A similar process of moving from directed attention through increasingly indirect prompts with a view to spontaneous behaviour by students applies to any of the study techniques and many of the other tactics suggested in previous chapters. The underlying mathematical action that students perform on their course material leads to them learning more effectively and more efficiently.

Scaffolding involves explicit action, while fading involves being less and less explicit and more and more implicit. Consequently, the tension between being explicit and being implicit is always present. Finding a way of working that is comfortable for you but at the same time effective for your students may mean altering your degree of explicitness somewhat, because although what really matters is what your students learn, what they want in the short term may not serve them well in the long term. This is yet another example of the didactic tension: what students want and expect may not necessarily be in their best long term interests, but if you choose to act against their apparent short term desires, you need to first gain their trust.

Motivation

Confidence and success go hand in hand, and are the best motivators. Each success (as long as it is seen as substantial and is valued by a teacher) boosts confidence, and confidence enables initiatives and risks to be taken which can lead to success. Each supports the other, and so to a large extent the two develop together as well. However, it is a rather delicate matter, as the tension between insufficient and excessive challenge indicated (see *Issue: Challenge – Insufficient or Excessive*, p166).

Issue: Responding to 'Why are We Doing this?'

When students ask, 'Why are we doing this?', they may be asking for examples of where the ideas and techniques could be used, particularly if they are taking mathematics as a service course. It is wise to be prepared for this, even to consult colleagues in other departments and to search the web in order to find good examples. A 'good example' is a problem that your students readily recognise as being pertinent to them, and whose solution they can follow once the techniques and theorems have been developed. However, there is another possible interpretation of the question that may be harder to deal with.

To investigate this, I interrogated my own experience: when do I tend to ask this sort of question? I found that I asked it when I was dissatisfied and frustrated, when I felt incompetent and unable to cope. I did not very often ask it when things were going well and I felt confident in myself. This led to the conjecture that perhaps my students were similar. It may be that what students are really saying is that they have lost their way and no longer feel they can do what is asked of them.

One way to test this conjecture is to train yourself to respond immediately to such a question with, 'It is absolutely vital in the steel industry.' (or some other setting of your choice). When I have done this, I have found my students' faces fall: this is not what they expected, and not what they wanted to know. Perhaps they even fear that I might start going through the details! In this way I convinced myself that often what students are doing when asking this type of question is asserting that they have lost their grip. Recognising this, I can use it as an entry point to find out where the difficulty lies.

It is certainly possible that a student is genuinely interested in how something so apparently abstract could ever be used, but their tone of voice, their posture and gestures, even the form of their question usually suggests whether or not they have a genuine interest.

Issue: Developing Facility

It is commonly believed that 'practice makes perfect', that students need to 'go home and do lots of examples'. Weaker students may indeed need to spend more time rehearsing techniques than their stronger colleagues in order to automate techniques (indeed this may define the 'strength' of students), but there is a catch: the more attention they give to practising a technique, the less likely they are to automate it! The reason is that 'facility' describes a state in which only minimal attention is required in order to carry out the procedure (Hewitt, 1996). The difference between an expert and a novice when carrying out a technique is that a novice attends to the details involved in each step, while an expert attends to the overall goal, to what happens next, and only peripherally to the details. (See *Tactic: Diverting Attention*, p85.)

If a novice puts enormous effort into doing a collection of exercises, they may become more familiar with the type of problem, but they may not gain facility with it. Their confidence may even be misplaced, for although some students are not satisfied until they think they could answer a similar question again in the future, many students are satisfied with simply getting answers to the set tasks. This, of course, is the *contrat didactique* again. The various tactics concerned with getting students to construct their own examples and problems are intended to stimulate them to get out of the rut of doing questions one after the other, and to stand back and try to make sense of the process, to become articulate about the class of problems they can tackle.

What about facility then? The best advice is to try to redirect your students' attention away from the details of a technique. This can be done by giving them something else to think about. In other words, set them something to explore, where they are led to construct examples for themselves and have to use the particular technique so as to gain information useful to the exploration. For example, to stimulate you students to work with the equations of straight lines, you could set them the task of finding the intersection of the line $y = 2x - 1$ with the quadratic $y = x^2 + x + 1$, and perhaps several more questions like it. This focuses their attention on the task of finding intersections. An alternative is to set them something more general: give them the parabola $y = x^2 + x + 1$, and ask them to find straight lines which intersect it in two distinct points. This way, your students have a larger goal, and are responsible for choosing their own examples and then generalising.

Hopefully part of their thinking will be taken up with using the answers in order to find something out, so they are 'doing examples' for their own purposes, not just so they can hand in the answers for homework. They probably want to do only enough examples to enable them to see what is going on, and probably as quickly as possible, and these are ideal conditions for automating a procedure and turning it into a skill. As well as giving them questions to explore, you can use the various tactics proposed in the earlier chapters.

Issue: Locating Surprise

Are there features of examples or problems that are inherently interesting and which arouse everyone's curiosity? It seems clear that no task, example, or problem is going to stimulate the curiosity of all students, even of all students who want to be mathematicians. People are different. More significantly, interest is not really a property of a problem, but of a person in a situation; it is the person who becomes interested in a problem, situation, or example. Thus, to stimulate the interest and curiosity of students, it is necessary to display curiosity oneself (without being overwhelming), and to create conditions in which students have the opportunity to be surprised and intrigued. It takes time for them to appreciate what is being offered, however; it cannot be hurried.

Behind most theorems there is some surprise or delight, even if it is only due to the tidiness of clarifying some issue. It may be that a theorem classifying some objects is a culmination and expression of intuition, but even so it is a response to a nagging question: could there be any other objects satisfying this property? For example, it is actually amazing that that all bases of a vector space have the same size; wonderful that the harmonic series does not converge but that any increase in the exponent makes it converge. In fact, most theorems are surprising or significant in some way, even if they simply confirm intuition or establish a technique (Moshovits-Hadar, 1988).

Locating and re-experiencing the surprise at or significance of something that has become routine and obvious, and using that to decide what to stress, can make a huge difference to students' appreciation of a topic. It can also help lecturers and tutors to display their interest and enthusiasm. Lecturers who are not enthusiastic about their topic turn their students off even more than lecturers who are over enthusiastic!

To locate the surprise, take a standard theorem, definition or technique you are about teach, and ask yourself what made this result remarkable to those who first struggled to express it. What question does it resolve, and what was it like not to know the answer to that question?

For example, if the series $\sum a_n$ is absolutely convergent then it is convergent. This apparently unremarkable result is confirmation than absolute convergence is stronger than convergence. Indeed, if it were not true, the definition would contradict ordinary English usage of adjectives, but it still needs to be confirmed mathematically. It is a formal demonstration that intuition is correct. Most students see it as unremarkable because if it were not true then no one would mention it. They often do not appreciate the organisation and interconnection of related ideas, and hence either do not pay attention to, or do not appreciate, the significance of a result. The best way to appreciate a result's significance is to see it used in a variety of contexts. The best way to appreciate the importance of conditions and assumptions is to construct examples, initially very general and then with increasing constraints, in order to show that a result is the 'best possible'. (See *Tactic: Boundary Examples*, p14, 136.)

In first year courses where students are seeing familiar mathematics treated more formally (calculus or analysis, for example) and at the same time are meeting new ideas introduced fairly formally (in linear algebra, for example), many become very confused as to what they are permitted to 'know', and what the overall enterprise is about. By rediscovering and then communicating your own surprise, you are likely to become more aware of the formalising process itself, and so be of explicit assistance to students.

Rather than focussing on answers to problems, and proofs of techniques, you can focus your attention on what students are making of a concept by asking them:

- to draw pictures or networks of concepts;
- to explain a concept to others;
- to 'write to learn' (see *Tactic: Advising Students How to Learn How to Learn (Learning Files)*, p98; *Advising Students How to Study a Mathematics Text*, p96);
- to devise a simple and a complex example illustrating a generality (see *Tactic: Expressing Generality*, p14) or demonstrating that they understand a technique or theorem;
- to devise an easy, a hard, and a general question of a given type (see *Tactic: Particular – Peculiar – General*, p88);
- to express a generality by constructing the most general problem they can of a given type;

Further examples can be found in Watson and Mason (1998).

One way to uncover the surprise in a theorem is again to seek its historical roots. What sorts of variations did mathematicians consider? How did the relevant definitions come about? What motivating problems and the informative examples were used by different mathematicians?

Despite being well worked out and 'finished', even the most apparently elementary mathematics can still be susceptible to development. Here are a few examples.

- While preparing the intermediate value theorem for presentation it occurred to me to ask what could be said about functions that satisfy the intermediate value theorem on every closed sub-interval of a given closed interval (known as the Darboux property; see Boas, 1960). Can such functions replace the assumption of continuity in other theorems?
- A standard theorem is that a continuous function on a closed interval achieves its maximum and minimum. Is the same true of an integrable function?
- If I have a collection of techniques such as different methods of integration, can I construct one problem that *requires* several of them, or even all of them, to be used, or the same technique to be used more than once?
- Students often struggle at first with the notion of a function specified by different rules on different intervals (glued functions), so can I construct some clever glued functions, perhaps ones that are glued at an infinite number of points but are nevertheless continuous or differentiable? For what sorts of theorems do glued functions provide boundary examples (by being k times differentiable but not $k+1$ times, for example)? Can you construct a function that is a polynomial of degree n on the interval $[1/n+1, 1/n]$ that is differentiable at least n times on the interval $[1/n, 1]$ and is well-defined at 0?

Even if the questions are not original and unsolved, they may be beyond what you recall of the topic. Working on these in relation to your lecturing (not to displace further research time) can add an element of discovery and freshness to rehearsal of the ordinary material.

Issue: Exploiting Surprise, Interest and Enthusiasm

When a student actually asks a question, it is very tempting to provide a quick answer. For example, Robert Baldino (email discussion) pointed out that when a physics student asks why L'Hôpital's rule works, an epsilon-delta-Weierstrass proof may not be very helpful. An approach based on Cauchy's mean-value theorem with hand waving may not be satisfactory to an analyst, but it may be more helpful in providing images that inform your students' appreciation of the technique. An intuitive sense of a concept is often more useful than a formal definition, especially to begin with, and it is also more likely to enrich the definition when one is needed.

Saying that a proof will come later (or in a later course) may satisfy once, but it is unsatisfactory as a habit because deferred gratification goes against a central feature of modern culture. It is also important to find out what the student is actually asking before launching into an explanation that would satisfy an expert (see *Issue: Responding to 'Why are We Doing This?*, p179). It has been suggested that students are often given 'explanations' (proofs) before they are ready, and so develop an antipathy for formal justifications that blocks their later progress.

Students can be surprised, and it is through examining their surprise that they can be intrigued into wanting to know more. This can be exploited, particularly by extrovert tutors who enjoy working their audience like a vaudeville performer. However, even those of us who are less inclined to this sort of performance can still achieve a great deal by locating the surprises that are part of the topics we teach (see *Issue: Locating Surprise*, p181). Even if you are not initially interested in a result, trying to recapture the original interest that made it important at one time can help your students to appreciate and become interested themselves. A tutor who is unenthusiastic is quite likely to display and communicate it through their posture, gestures, and tones of voice. On the other hand, a tutor who is wildly enthusiastic is likely to be so caught up in the ideas that their speed of delivery and general energy will put at least some students off. However, there is a wide domain of behaviours between these two extremes that is well worth exploiting.

Issue: Intriguing and Pointing the Way

The standard mechanism for motivating students is to start a new topic with a discussion of some typical problems that the topic can resolve, perhaps with applications in domains of particular relevance to the students. Sometimes tutors then include some examples at the end of the topic by way of application, but there is much more to applying techniques than this. As suggested earlier, you can know what to do, and even have considerable facility in doing it, yet not recognise an opportunity to use a technique.

Resources Upon Which to Call

You cannot do your students learning for them: they must do it for themselves. However, you *can* make use of the many resources you have to hand. These include your students' own powers and experience, your awareness of common mathematical themes which keep recurring and which can serve as connectors between otherwise apparently disparate topics or techniques, and of course your own experience. This section develops these notions a little further in order to highlight the ideas that underpin the tactics proposed in the previous chapters.

Theme: Mathematical Powers

Seeing students as sponges ready to soak up mathematics does not really do justice to them as sense-making human beings with desires and fears, interests and dislikes. Seeing them instead as possessing powers that may need to be brought out into the open and developed offers a more fruitful perspective for thinking about teaching and learning. For example, every student will already have displayed the power to specialise a generality and to detect general patterns, but they may or may not have honed it within the context of advanced mathematics. If their lecturer always does the specialising ('Here are the important examples.') and the generalising ('The general result is …') then students may get the impression that they are not expected to use those powers for themselves. As a consequence, these powers may be put to one side, and they may be reinforced in being dependent upon their lecturer. By contrast, explicitly calling upon students to take the initiative, as suggested in the many of the tactics in the previous chapters, engages students in the process of mathematical thinking by making them aware of their powers.

Similarly, all students have experience of forming mental images, and of expressing their thoughts in diagrams, words and symbols. However, if they get the impression that it has all been done and that their job is simply to 'learn' what others have said, then they are unlikely to make much progress in mathematics. By calling upon your students to use their powers of mental imagery, and by explicitly using your own ('I see things this way ...', or, 'Imagine a ...', as suggested in several of the tactics) you can remind them of the importance of using those powers, and show that you value them and use them as well. Mental imagery need not mean pictures, indeed for a substantial number of people it can means sounds, or even a more vague sense that is imprecise but informative, and perhaps not too distinct from intuition.

All students have also formulated conjectures and tried to convince themselves and others of the validity of those conjectures. However, lectures can sometimes give the impression that mathematics is entirely worked out and unproblematic, because the conjectures, the modifications, and the wrong turnings are erased from the record. By exposing your students to a conjecturing atmosphere, by encouraging them to make and modify conjectures, and to try to convince first themselves, then a friend, then a reasonable sceptic, you can support them in thinking mathematically. It is as important to learn to be mathematically sceptical as it is to be good at generalising. It is often possible to praise people for making a conjecture, and especially for modifying a conjecture, without needing to praise the content of the conjecture itself.

Another natural activity that all students have done before is ordering, organising, and classifying. Again, however, they may not be aware of how central these skills are to mathematics. Deciding which of two conditions is stronger and which weaker, sorting out the implications of 'if and only if', appreciating a proof of the equivalence of several different hypotheses, and appreciating the effects on illustrative examples of adding or removing constraints or conditions, are all examples of mathematical ordering and organising. Moving from discerning properties to characterising the objects that have those properties is part of the process of axiomatisation. Many theorems can be thought of as providing alternative characterisations of how to recognise an object with a specific property (such as continuous and integrable functions, finitely generated abelian groups, or normal subgroups).

The tactics presented in the earlier chapters are designed to assist with the process of stimulating students to take the initiative and to use their own powers, rather than depending on you to do the work for them. In preparing to teach any topic, it is worth taking a few moments to identify illustrative examples that you could use with your students to stimulate them to abstract their essence, to generalise and perhaps even conjecture some of the theorems. More suggestions about how to prepare for your teaching are given in the last section of this chapter.

Theme: Shift of Attention

It is a common experience when teaching to find that, no matter how clearly you try to formulate and express an idea, there is usually at least one student who gets the wrong end of the stick, who hears or construes something different. This leads to the question of what it is that students are attending to when they are in a lecture, reading a text, or trying to do an exercise. As an expert, it is natural to think that your students see the same things that you see, that they stress the same features that you stress. Experience suggests, however, that students often stress different features. They are often blissfully unaware of the generality when their tutor is working through an example (see *Tactic: Making an Example be Exemplary*, p29, and *Issue: What is Exemplary About an Example*, p173).

Pondering the question of what my students attend to, in contrast to what I would like them to attend to, leads me to the question of what I attend to. If I do not know where my attention is focused, if I am not aware that I am internally stressing some features and ignoring others, then I will be unable to direct student attention appropriately.

For example, if I confidently treat functions, transformations, or functors as objects, while my students see them as processes to apply to objects, then there is a significant mismatch that will lead to trouble. The mathematical experience of turning processes and transformations into objects has been studied in some detail, both at school level (Gray and Tall, 1994), and at tertiary level (Sfard, 1991, Dubinsky, 1994). It is a non-trivial act. Similarly, if I use the numbers in an example simply as place holders while my students treat them as an integral part of the activity, then they will not gain the experience that I think I am giving them, namely of how to do all questions of this type. Similarly, when I announce a definition I am usually aware of why it is formulated in the way that it is, and what the consequences will be in terms of being able to state subsequent theorems succinctly. A definition signals a focus of attention, and most definitions have been refined over a period of time.

These considerations led to the following conjectures:

- Each technical term in mathematics signals a shift in perception, in the focus of attention, experienced by those who formulated or refined the definition.
- Each student has to experience a corresponding shift of attention in order to use the term meaningfully and with facility.

The only way to investigate the validity of these conjectures is to examine your own experience. In mathematics education the point is not to try to prove or refute these conjectures, but rather to examine whether they can help to inform future practice by offering something to think about when teaching. For example, these conjectures might lead you to try to re-experience the shift from not having a term at your disposal to having-it, in order to see what might be involved for your students.

For example, the ideas of area and integration seem reasonable to entering undergraduates, but are seen quite differently by mathematicians aware of some of the difficulties and of the definitions and theorems that clarify the ideas and make them precise. Students are at first perplexed as to why there is so much fuss, and then become confused by the intricacy of the definitions, first of Riemann and then Lebesgue integration. Similarly, students entering university are convinced intuitively that $\sqrt{2}\times\sqrt{3}=\sqrt{6}$, and until they experience the problems experienced by Dedekind, they are unlikely to appreciate Dedekind's construction of the reals, or anyone else's.

I have already suggested that drawing your own diagram allows you to make much more sense of the necessary and the arbitrary, the invariances and the constants, than looking at someone else's completed diagram. The same applies to the proof of a theorem or to a worked example. Whenever you read some mathematics, or write it, you are stressing some aspects and ignoring or downplaying others. The task of a teacher is to communicate this focusing of attention to their students without being too overt or over-bearing. This can be done by the use of pauses, modulated voice tones, changes of pace, and revealing your own enthusiasms.

Frameworks for Informing Teaching

This section offers three frameworks on which to build, or with which to structure, preparation for teaching. Each makes distinctions that have been found to be helpful.

Framework: Manipulating – Getting-a-sense-of – Articulating

In discussing the issue of doing and construing, mention was made of students making sense of ideas. Indeed, the whole thrust of this book has been to put forward ways of stimulating students to take the initiative. The expression 'make sense' is interesting because it summarises one of the principal approaches to teaching since Plato, if not before: try to make things relate to students' sensations by giving them objects to manipulate. The link between handling objects and making sense is through sensation. By the time students get to college and university we assume that 'making sense' is a more abstract or virtual process in that the objects they manipulate are more symbolic than physical. The principle still remains that students want, indeed need, confidence-inspiring familiar objects to manipulate and on which to try out new ideas so that they can literally 'make sense' of them.

It is tempting, as an expert, to think that students ought to pick up ideas at the first or second encounter. After all, they seem so simple and logical! However, most people do not master ideas on their first encounter. Instead we have to 'see an idea go by', and only after renewed contact, during which we gain confidence and familiarity and make connections, do we approach mastery.

It follows that in order to understand something, it helps most students to have some familiar, confidence-inspiring examples to manipulate, through which they can experience the general idea. As they manipulate these examples, they can begin to get a sense of what is going on (what is exemplary, and what is being exemplified), and over time this 'sense' can be expressed more and more articulately. Eventually the articulation will become fluent, and can then be used as a confidence-inspiring component for further sense-making. Thus it is absurd to define a 'quasi-inverted partial thump' in terms of 'semi-straight bloboids' unless the latter is already familiar as a concept. It may be a trivial observation, but it is frequently ignored. Students are often introduced to several concepts in succession, each depending on the ones that have gone before, and assumed to have mastered each on their first exposure. This spiralling process can usefully be seen as a helix.

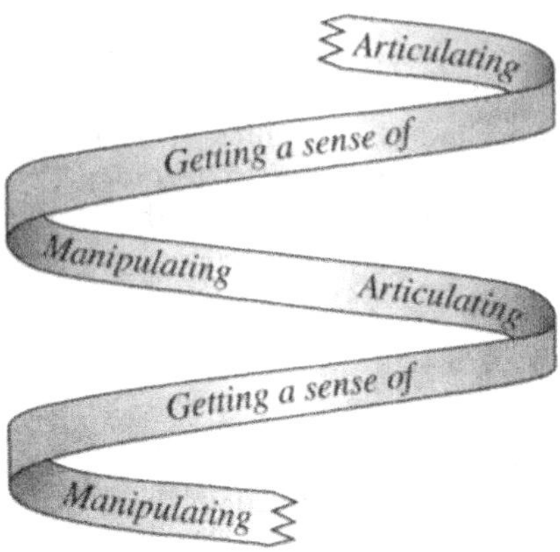

When difficulty arises, it is possible to retreat back down the helix, or even to leap down, and then to rebuild confidence and understanding while working your way back up again.

It also takes time for students to become confident with using unfamiliar symbols and terms to express their own thoughts and ideas. When people meet a succinct notation for a complex idea (whether a mathematical one, or an educational one like *tactic* or *framework*), they spend some time (perhaps very short, perhaps not) carrying the expanded meaning along with the succinct. For some time, when they encounter the succinct version, they find it convenient to say to themselves, even to write down, the expanded meaning. Over time, as they gain familiarity and confidence, the succinct notation begins to trigger relevant connections and meanings automatically. For example, on first encountering 'linearly dependent', 'Hausdorff', or 'random variable', students are likely to need access to the definition, to a rich example, to images or pictures, and to relevant formulae. As this 'concept image' becomes more secure, more confidence-inspiring, the succinct will begin to truly symbolise and trigger these rich connections as required.

For example, Michener (1978), in looking at what constitutes understanding for mathematicians, developed the notion of examples-space, results-space and concepts-space as components of understanding. Mathematicians' attention shifts easily from one to another, so to capture this she introduced the notion of links between these spaces. She made a large number of distinctions in trying to develop a representation of components of understanding. Whether all these distinctions are useful for students in their attempts to 'learn' is open to question. However, a lecturer or tutor who is aware of different statuses and roles for examples, results, and concepts is in a much better position to emphasise and de-emphasise, to focus attention or not, in order to help their students make sense of the ideas.

First, Michener distinguished several types of examples:

- *Startup examples*, which are used to arouse interest and to suggest how the theory would develop;
- *Reference examples*, which are mastered and used to test future conjectures or to revisit concepts;
- *Model examples*, which are paradigmatic and generic (see also Mason and Pimm, 1984; Rowland, 1998);

- *Counter-examples*, which demonstrate the boundaries of concepts or techniques, or which contribute to counteracting standard misconceptions.

Secondly, she distinguished several types of results:

- *Basic results* establish elementary but important properties of concepts and examples (that the eigenvalues of a matrix are related to the determinant, for example) linking procedures with definitions;
- *Key results* establish fundamental and repeatedly used facts;
- *Culminating results* are landmarks that provide both motivation and a resting point for students;
- *Transitional results* are lemmas and theorems that are needed for completeness or as technical items with limited use.

Finally, she distinguished several types of concepts:

- *Definitions* (intensive and extensive, see p25);
- *Heuristics*, in the form, 'Try working backwards from what you want to find.', or, 'Try a simpler case or situation to see what calculations you need to do.', or, 'If someone told you the answer, how would you check it? Now use the same calculations with letters for the quantities that are as yet unknowns.';
- *Mega principles*, such as, 'look at extreme cases', 'symmetric matrices are nice', or 'try special cases involving 0s and 1s';
- *Counter-principles*, such as, 'watch out for division by 0', 'watch out for multiple roots', or, 'when changing variable of integration, don't forget to calculate the new differential and the new limits'.

Michener advocated immersing students in the practice of being asked (and hence asking themselves) broad questions based on these distinctions: 'What kind of an example is this?'; 'What sort of a result is this?'; 'What kind of concept am I being offered here?'.

Framework: See – Experience – Master

Rough distinctions can be made between three types or levels of engagement with new ideas, which can help you to keep in mind the fact that it usually takes time and multiple exposures before new and difficult ideas become familiar tools with which to think. We can distinguish between

- *Seeing* an idea go by, a student's first encounter with it, and first contact with its particulars;
- *Experiencing* an idea through multiple exposure and beginning to appreciate a generality;
- *Mastery* of facility and confidence, and ability to articulate the general.

These distinctions (which are certainly not intended to be hard or clean) may be illuminated by the simile of standing in a train station. When an express train roars through, you have a sense of power and speed and of its rough shape. This initial encounter is like meeting a new idea, when there is a similar sense of 'seeing something rush by'. When freight or non-stopping local passenger trains then go through more slowly, you can see some of the detail, get a sense of its nature and role. This corresponds to having multiple exposures to ideas, perhaps in several contexts. Finally, when a passenger train comes through and stops, you can get on. This corresponds to developing facility, confidence, and detailed experience; you 'climb aboard'.

Some pedagogies try to use only the stopping passenger train, intending students to comprehend a significant amount on their first encounter. There are several weaknesses to this approach. The first is that it ignores the psychological and social experience of a student encountering something new. Difficult new ideas always seem to rush by, no matter how invitations their tutor makes to 'climb aboard'. It all seems frantic and confusing. Other pedagogies are content to allow frequent exposure to express and freight trains without the passenger train ever stopping. The result then is that students never have a chance to rework ideas and techniques for themselves in the headlong rush to cover the curriculum.

When students encounter a new idea (limits, continuity, topology, group, ring, or vector space, for example), they try to relate it to what is already familiar. This, of course, is the role of examples (see *Framework: Manipulating – Getting-a-sense-of – Articulating*, p187). However, on their first encounter most students need time for ideas to mature, to percolate, to connect, to become their own. For this reason, it can be useful to carry the meanings of new symbols forward for a while (perhaps on a second OHP or board) until students can be weaned off the support (see *Tactic: Directed – Prompted – Spontaneous*, p91).

Framework: Awareness, Motivation, and Behaviour

So far in this chapter I have mentioned the mathematical powers that students possess and mentioned the issue of explicitness. If you try to be too subtle about stimulating students to use their powers and about the presence of mathematical themes, about the role of definitions and the nature of proof, then it is very likely that it will all pass most students by. Yet one way of looking at the teaching of mathematics is that you are trying to awaken and extend your students' awareness of the nature of mathematical thinking as well as training them in certain behaviours or techniques. This can be done by harnessing their emotions, which is what motivation is really about. This three ideas of educating awareness, training behaviour, and harnessing emotion, when elaborated, become a useful structure or framework around which to organise the teaching of any topic.

Framework: Concept Images

Concepts, or mathematical ideas or notions, are not just abstractions. They are more like nodes with connections to mental images, past experiences, techniques, and typical problems. The notion of a 'concept image' (Tall and Vinner, 1981) tries to capture what it means to have a sense of a concept. It was developed through research which showed that what comes to a student's mind when they hear a technical term is rarely what they were taught, and certainly not what comes to their lecturer's mind.

It is useful to break the notion of a concept image into three interwoven dimensions corresponding to aspects of the psyche (cognition or awareness, emotion, and behaviour or enaction):

- root or source problems and contexts in which the concepts or topics reappear (emotion);
- familiar and new language (definitions, for example) and techniques associated with the concepts (behaviour);
- images, connections, and sense of the concepts, together with standard confusions, errors or misconstruals students have or make (awareness).

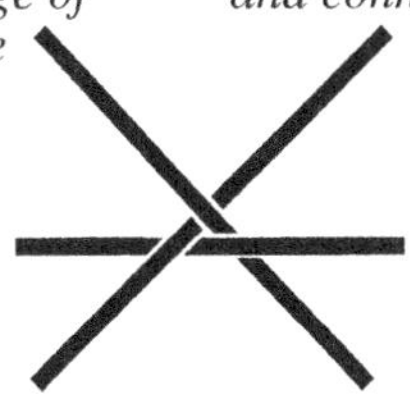

This three-strand structure can be used as a framework when you are preparing to teach any topic. The motivational-emotional strand suggests that it is useful to remind yourself of (and so be in a position to indicate to students) the original problems whose solution gave rise to this topic. For example, one stimulus for probability came from solving the problem of distributing gambling stakes fairly when a session was interrupted; integration originated in finding areas; groups began as symmetries of solutions to differential and polynomial equations, as well as symmetries of physical objects; the scalar product began as a measure of length. It is useful to consider also the range of contexts and problem types in which the same idea has already turned up. This provides motivation for students (but does not necessarily need to be used in the introduction to the topic). Do not forget that, for students, the only important questions are often types they will be asked to do in the assessment.

It is also useful to remind yourself of your own mental images, diagrams, idiosyncratic ways of thinking, and connections with other topics. At the same time, refresh your memory of the sorts of things that students struggle over or mis-construe, and see if you can find what it is about what you usually say or do that causes the problem.

Remind yourself, too, of the language that accompanies the topic and which may not yet be familiar to your students. This includes both words they already know but have not yet come across in their technical sense, and terms they have not yet met (see *Tactic: Introducing Symbols,* p90; see *Framework: Manipulating – Getting-a-sense-of – Articulating,* p187). Finally, remind yourself of the techniques which they will have to carry out, and the inner thoughts that you have when you use them. Unless you are explicit about these, your students may not find out about them! (See *Tactic: Exposing Inner Monologues,* p57.)

Mathematical Themes

Doing mathematics involves handling a number of innately mathematical ways of thinking and developing an idea. Four themes are mentioned here, with the implication that it is worth drawing attention to them explicitly since they provide a structure for making connections between otherwise apparently disparate topics. When you become aware of one of them in a lecture or tutorial, you can choose to emphasise it explicitly, or perhaps implicitly, so as to help students see connections across topics.

Theme: Invariance Amid Change

Invariance is a major component of many mathematical theorems, as well as capturing the essence of most mathematical properties. For example, distance and area remain invariant under rotation about a point; the angle sum of a planar triangle remains invariant under change in shape; all scalar products are invariant under change of coordinate system; shears preserve area; absolute convergence of a sequence is preserved under term-by-term decrease in absolute value. Unfortunately, the statements of many theorems stress the invariance but underplay the range of change within which the invariance holds. Many students therefore fail to appreciate the implicit dynamic and inherent generality in theorems.

Invariance amid change is the central issue for students when they are uncertain as to which aspects of an 'example' are permitted to change, and which aspects are structural and invariant. (See *Issue: What is Exemplary About an Example?*, p173.)

Here are some common examples.

- Students meeting partial differentiation for the first time often feel it is magical. They are uncertain as to when things should be treated as constant and when they have to be differentiated, and why the procedure works.
- Students meeting epsilon-delta proofs are often unclear about which is to be chosen and which is, in effect, chosen by someone else and not under their control.
- Students presented with a technique in linear algebra such as row-reduction are often not clear why some operations are permitted and others not, and so they are led to perform calculations incorrectly.

Techniques that are mastered without understanding what the technique is doing are bound to have a magical feel about them. Students are rarely able to immediately take on board someone else's complex story about how and why a technique works, but they can be supported in piecing together their own story by working on generic examples (see *Tactic: Story Telling*, p82).

Theme: Doing and Undoing

Factoring polynomials is the undoing (opposite) of expanding brackets. Partial fraction decomposition is the undoing of adding algebraic fractions. Differentiation and integration are the undoing of each other. Solving a differential equation is the undoing of differentiating a function and looking for relationships. Solving linear equations is the undoing of substituting values into expressions. Discovering empirically that the rate of diffusion of water through uniform sand is directly proportional to the length of the path and inversely proportional to the head is one problem, but deducing the transmissivity of inaccessible rock and soil from head and flow data obtained from a test well is the inverse problem. Tomography is also an inverse problem, as is using the natural frequencies of a vibrating system of masses and springs to deduce those masses and spring stiffnesses. (See Groetsch, 1999, p13.)

Many important mathematical theorems concern the (usually creative) act of solving the undoing problem, 'If this were the answer to a problem using that technique, then what might the question have been?'. Drawing this theme to the attention of students can help them to make connections and develop an overview through encountering the same idea in different contexts. It can also help them to appreciate the kinds of questions that mathematicians ask themselves which lead to fruitful exploration and investigation. (See *Tactic: Doing and Undoing*, p89; see Groetsch, 1999, for a wide range of examples.)

Theme: Freedom and Constraint

An equation or inequality is a constraint placed on a collection of free variables. A differential equation imposes a constraint on an otherwise general (adequately differentiable) function, and can often be usefully viewed as a function on a set of functions. A random variable is a value that is constrained by some probability distribution, and can usefully be seen as a function. A theorem states conditions (constraints) that must be imposed on an otherwise free or general object. A geometrical construction problem imposes constraints on otherwise free or general geometrical objects. Much of mathematics can be seen as the study of sets of constraints, addressing the question of whether there are objects meeting all those constraints and, if so, how to find some or all of them. (See *Tactic: Boundary Examples*, p14, 136)

Theme: Extending Meaning

At one time, the only numbers that were thought to be sensible were whole numbers. Operations on whole numbers were then extended to negative numbers, fractions, decimals or reals, and finally to functions, including polynomials, and operators such as differentiation and integration. Seeing these as extensions of the original concept is necessary in order to justify the use of the same sign (+, –, ×, /), and to appreciate their mathematical structure. Sums were extended to integrals, then to line integrals, contour integrals, and on to higher dimensions. Arithmetic and algebra with single objects was extended to Cartesian pairs, then to higher dimensions. The notion of a sequence was extended to include functions on the positive integers, then to nets. The notion of an open interval on the real line was extended to an open set in topology.

Each stage of extension required letting go of some of the implicit assumptions about the objects or about the specific meaning. This meant letting go of something concrete and confidence-inspiring, and accepting a more abstract property or condition as fundamental. The change to considering more than three dimensions is a prime example of how attention and confidence has to shift from pictures and mental images to algebraic work with coordinates. The images have to become more abstracted, less specifically three dimensional. Many students never really accomplish this shift. More could, however, if they were assisted by experts who were more explicit about the process, more attentive to their students' struggles, and more reflective about their own experience.

By being aware of the role that extension plays in the development of a topic you can at least have sympathy for your students, and you may be able to make helpful remarks about where you direct your attention when you think about the ideas involved.

Theme: Organising and Characterising

The desire to organise and classify things is widespread, but many students do not appear to realise how strongly it drives mathematicians. The wish to reduce complexity to a few basic principles, to be certain that there are no other objects lurking about other than the ones already discovered, is very strong. The desire to organise is not necessarily satisfied by encountering someone else's neat classification scheme. Indeed, someone else's scheme is more likely to act as spur for you to try to find an alternative. However, the collections of theorems that constitute a topic can be viewed as the identification of one or more classes of objects, with different types of objects put into the same classification scheme, and the classification as constituting an organisation of the material. Once everything is connected up, mathematicians lose interest, and students are often subjected to the final result but not to the classification process. Encountering a theorem that proves two or more properties or axiom systems to be equivalent can produce excitement amongst students, if it is presented in a way that stimulates their own desires, but it can also become a deadly game of logic, chasing if and only if statements round in circles.

Reflection

Task: Reframing

Consider some aspect of your current teaching, and try to cast it in terms of these (or other) tensions, actions, and frameworks.

Appendix A: Exploratory Tasks

Despite the frequently made claim that students do not have time to explore mathematics deeply, it seems odd to claim to have taught someone a mathematical topic and yet not have had them tussle with challenging questions in that domain. This appendix offers some sample questions that could be used as starting points for such an exploration. The intention is that any student who successfully works through to an answer has demonstrated a good appreciation of the topic.

- Given a specific non-singular linear transformation from R^3 to itself, which three by three matrices can represent it?
- Discuss the mathematical methods needed to analyse the situation shown in the left-hand diagram, where the spheres are not necessarily small, the 'string' has mass and is extensible, the pulley has friction, and the table is not smooth.

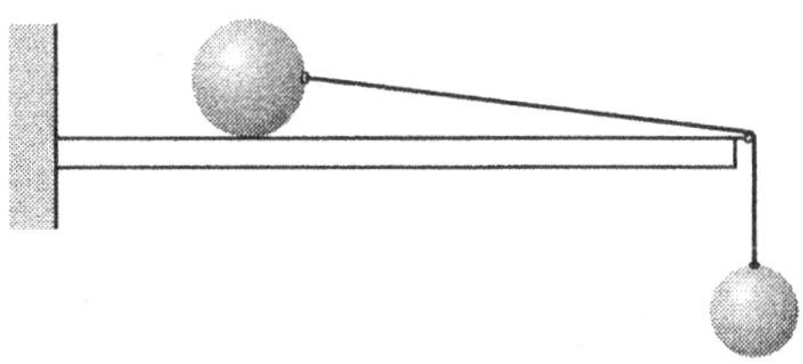

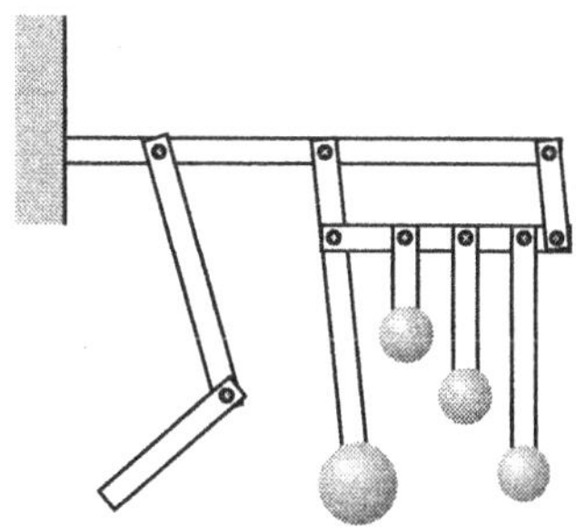

- Given a compound pendulum as illustrated on the left of the right hand diagram, what will be the result of displacing the bottom bar only, the top bar only, or both bars? Given a compound pendulum as shown on the right of the diagram, arranged so that the bar oscillates following the large pendulum, what effect will displacing the large mass have on the motions of the small ones? (The computer software Interactive Physics makes these sorts of situations easy to display, both as support for developing intuitions, and as stimulus for mathematical modelling.)
- Which sets of real numbers can be the set of cluster points of some set of real numbers?
- Which real-valued functions can be the answer to an integral that can be found by integrating by parts n times, for $n = 1, 2, \ldots$?
- Must the sum of two periodic functions be itself periodic? Is the product likely to be periodic, and under what conditions will it be?
- What exactly is a point of inflection? Does $f(x) = x^4$ have one, and how can you tell algebraically?
- What exactly is an asymptote? What conditions guarantee that there will be an asymptote?
- Which numbers, N, have the property that every power of N can be written as the sum of two squares? (See Wisner, 1999)

- For which values of x do the expressions
$\sqrt{1+x\sqrt{1+x\sqrt{1+\cdots}}}$ and $2+\cfrac{x^2-1}{2+\cfrac{x^2-1}{2+\cdots}}$ take finite, real values?

- Everyone knows that $f' = f$, with suitable conditions, can only be solved by the exponential function. What about $f' = f^{-1}$ and, more generally, $f' = f^{(n)}$ or $f' = f^n$ for various values of n? (See AMM problem E2105, 1976, p696)

- I have made the conjecture that if a function f is differentiable at a, then the limits of the derivative of f as $x \to a^+$ and $x \to a^-$ will be the derivative of f at a. Use the function $x^2 \sin(1/x)$ $(x \neq 0)$ to construct a class of counter-examples. What is it about your functions that makes them work as counter-examples and how might they be generalised?

- Let $g(x) = \begin{cases} 0 & 0 \le x < 1/3 \\ 1 & 1/3 \le x \le 2/3 \\ 0 & 2/3 < x \le 1 \end{cases}$. Discuss whether it is possible to

  integrate the functions specified by $f_1(x) = g(x)$, and

  $$f_{n+1}(x) = \begin{cases} 1/2\, f_n(x/3) & \text{if } f_n(x) = 0 \\ f_n(x) & \text{otherwise} \end{cases}$$

  on the interval $[0,1]$, or the function $f(x) = \lim_{n\to\infty} f_n(x)$.

- If r is less than or equal to the radius of curvature at each point of a curve that is twice differentiable, can a circle of radius r roll along the curve without getting stuck?

- Which subsets of the power-set of a finite group are themselves subgroups under element-wise multiplication of sets?

These exploratory problems should not be confused with any project work that you assign to your students, where the purpose is to get them to read an exposition of some mathematical ideas, work through it, and then re-express the ideas so that their colleagues can understand them. Here, the aim is to promote mathematical thinking and to encounter several ideas associated with a topic, while also having to select what seems relevant and appropriate from a range of possibilities.

Appendix B: Convergence Case Study

This topic is usually the subject of only two or three lectures and associated tutorial sessions. Consequently, the range of tasks suggested here may appear to be excessive and impossible to implement. However, it is only by reviewing a collection such as this you can make a sensible choice to meet the needs of your particular students. Furthermore, as is argued elsewhere (see *Issues: Time – Coverage and Pace*, p164), if students feel they have mastered a topic, and in addition have learned how to learn mathematics, they are likely to be more effective at tackling further topics more efficiently than if they feel that topic after topic is going by in an unstoppable rush.

The structure of this appendix exemplifies the framework *Preparing To Teach a Topic*, which applies to any mathematical topic. It is important to note, however, that the separate subsections were *not* constructed in the order in which they are presented here. Rather, different ideas emerged, and these triggered new entries in different places. The whole point of a framework is to use it to inform, not stifle, thinking. Using it mechanically can only lead to mechanical results.

Once a structured approach to preparing a topic has been initiated, further refinements can be added at any time as a result of your students' reactions and responses or when you chance upon something relevant. The notes you have made can be perused before subsequent presentations as a quick reminder of details that might otherwise be forgotten.

This appendix was constructed collaboratively with Bob Burn, for whose insight and experience I am grateful. Any remaining imperfections are, of course, my responsibility.

The underlying framework is that of *Concept Images* (p190), which requires that we begin with a list of salient items. These naturally contribute to aspects of the framework, and the framework can trigger questions that may need some research to answer. In this case, we decided that the salient items were as follows.

Salient Items

- Partial sums are a way of moving from a series to a sequence.
- The definition of the convergence of a sequence.
- Convergence only makes sense in the context of the completeness of the set of real numbers. Thus convergence of sequences of rationals is of little interest if you are confined to the rationals!
- Convergence of sequences is preserved under finite sums, scalar multiples, deletion of an initial segment, and deletion of finitely many terms. All these notions extend to series.
- Builds on geometric progressions (in a big way), partial fractions (a little), and induction (rather rarely).
- Positive version of d'Alembert's ratio test: $a_{n+1}/a_n \to L < 1$ as $n \to \infty$ implies convergence.

Though it is tempting to move quickly on to a more general discussion of arbitrary series, it is useful to introduce the idea of convergence of series through sequences with only positive terms. The generalisation can be made in the proper place. Focusing on positive series is likely to make the transition to general series more efficient and effective.

- Cauchy's n^{th} root test: $\sqrt[n]{a_n} \to L < 1$ as $n \to \infty$ implies convergence; useful for calculating the Cauchy–Hadamard radius of convergence.
- If d'Alembert's test works then Cauchy's test works, but not vice versa. (Take a series on which d'Alembert's test works, such as $\sum 2^{-n}$, and interchange the $2n^{\text{th}}$ and the $(2n-1)^{\text{th}}$ terms: Cauchy's test still works, but d'Alembert's test has ratios alternately greater and less than 1.)
- Comparison test: for two series, $\sum a_n$ and $\sum b_n$, if a_n/b_n is bounded above or below, then both series converge or diverge together; if $a_n/b_n \to L \neq 0$ then both series converge or diverge together.
- If f is decreasing and non-negative, then $\int_1^\infty f(x)\,dx$ exists iff $\sum_1^\infty f(n)$ exists.
- If f is positive and decreasing, then $\sum f(n)$ is convergent iff $\int\limits_1^{n+1} f(x)\,dx - \sum\limits_1^n f(k)$ is always convergent. (To demonstrate this, apply it to $\sum \frac{1}{n^\alpha}$ for $0 < \alpha < 1$.
- If f is integrable, then $\int\limits_1^{n+1} f(x)\,dx - \sum\limits_1^n f(k)$ is always convergent, even if $\int\limits_1^{n+1} f(x)\,dx$ is an unbounded sequence (for example $f(x) = 1/x^\alpha$ for $0 < \alpha \leq 1$, generalising Euler's constant).

Great examples

- $\sum 1/n$ and $\sum 1/n^2$ are critical examples that show how neither d'Alembert's ratio test nor Cauchy's root test are strong enough to deal with all cases. To appreciate convergence, it helps to be exposed to several methods for showing the first series to be divergent and the second to be convergent.
- $\sum \frac{1}{n^\alpha}$, which diverges when $\alpha \leq 1$ and converges when $\alpha > 1$.
- $\sum \frac{1}{n \log n}$, whose divergence can be proved using Abel's integral test. This arose as a response to Olivier (1826), who wrongly claimed that the series $\sum a_n$ is convergent if it has only positive terms

and $na_n \to 0$ as $n \to \infty$. (There is a beautiful account of this in Goar, 1999.) Abel (1826) then announced, '... and I can make more'. This provides a lovely snapshot of the role of conjectures and counter-examples in mathematical thinking. Abel further proved that no condition of the kind $F(n) \cdot a_n \to 0$ as $n \to \infty$ is sufficient to guarantee the convergence of the series $\sum a_n$.

- $\sum \frac{1}{n^\alpha \log n}$, which diverges for $\alpha \le 1$.

What do these examples exemplify? They could form part of an exploration of where the boundary between divergence and convergence lies. Could your students be encouraged to engage in an exploration based on this idea?

- $\sum \frac{1}{n} + \frac{1}{n^2}$ and $\sum \frac{1}{n} - \frac{1}{n^2}$ both have one convergent part and one divergent part. The first series is bounded below by a divergent series so is divergent itself; although the sum of two convergent series is convergent, if one series is divergent, the sum cannot be convergent, at least for positive series. The second series might appear to have a better chance of being convergent, but the terms become arbitrarily close to a divergent series, so it is still divergent.
- (A 'bread and butter' example.) Consider the tiling of a one-way infinite strip coloured as shown.

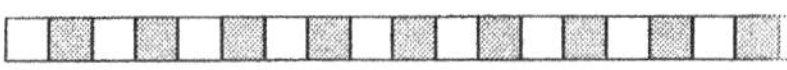

Here grey and white squares alternate. 'Clearly' the ratio of grey to white in the whole strip should be 1 : 1. However, consider grouping the squares into a set of disjoint 'tiles' in which tile *k* consists of the kth white square and both the $(2k-1)^{\text{th}}$ and $2k^{\text{th}}$ grey squares. Satisfy yourself that these 'tiles' can account for all the squares on the strip, but that the ratio of grey to white appears to be 2 : 1. Show that, by suitable specification of these 'tiles', any ratio of the form $p : q$, with p and q positive integers, can be obtained as the ratio of grey to white.

Prove or disprove the following generalisation. Given two divergent series of positive terms, $\{s_j\}$ and $\{t_n\}$, and a positive real number ρ, it is possible to group the two series as $\{s_1 + \cdots + s_{j_1}, s_1 + \cdots + s_{j_2}, \ldots\}$ and $\{t_1 + \cdots + t_{k_1}, t_1 + \cdots + t_{k_2}, \ldots\}$ in such a way that the sequence of ratios of corresponding terms approaches ρ. In symbols, this can be written as

$$\lim_{n\to\infty} \sum_{j=1}^{j_n} s_j \Big/ \sum_{k=1}^{k_n} t_k = \rho .$$

Concept Image Framework

The general framework for probing the different aspects of a concept was used in this instance, so a version of this framework, with the relevant details included, is shown below. Reading someone else's chart is akin to trying to read their notes: it is of necessity succinct and laconic, but it does show how such a framework can be used to inform the different kinds of questions that can usefully be asked.

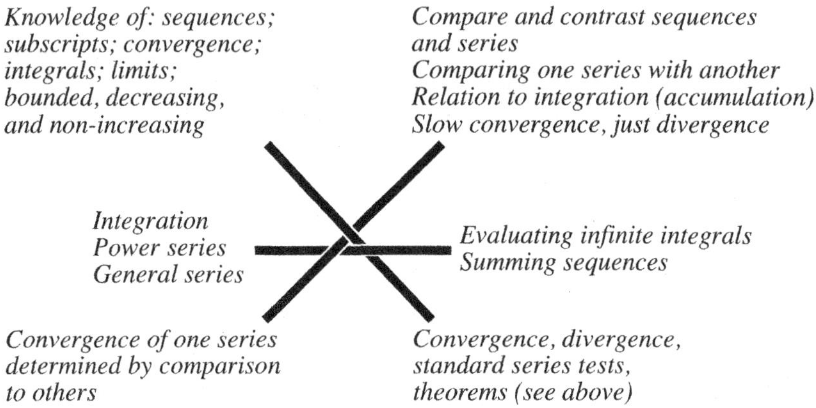

Keep a list of the confusions that your students develop, and use them to construct examples and guide them in constructing examples that challenge those misconceptions. For example, Abel's examples using logs were constructed in order to challenge Olivier's conjecture.

A firm grounding in positive series will hopefully lead your students to appreciate counter-examples to extending these ideas too simply to series of arbitrary terms, and to appreciate the role of absolute values and uniform convergence in the more general case.

Here are some possibilities for tasks to set students at different times.

Reconstruction	Make a sensible sentence which uses the terms .... Make a sensible sentence which links the two terms ... and .... State in words the roles of d'Alembert's ratio test and Cauchy's root test in studying convergence. Explain, as if to an intelligent but un-mathematical relative (who understands some mathematics but not this topic), what convergence of a series of positive terms means, and how to demonstrate that a specified series converges.
Exercises	Do as many of the exercises in the text as you feel you need to in order to tell someone when the ratio test might be useful, how to apply it, and what it does. Make up an easy and a hard question involving the convergence of a series of positive terms. Make up a collection of increasingly complicated examples of series that are convergent, but which converge ever more slowly, and with a slight adjustment can become divergent series. Make up and do a question which demonstrates that you know what convergence of a series of positive terms is about. Use the ratio test to show that the series $\sum_1^{\infty} \frac{n!}{n^n}$ converges. Generalise this result to exponents of the form $n \pm k$ in the denominator.
Exploration	Given a series of positive terms that converges, explore the extent to which the values of L_A and L_C can differ, where $\frac{a_{n+1}}{a_n} \to L_A$ and $\sqrt[n]{a_n} \to L_C$. Construct a series, $\sum f(n)$, for which $f(x)$ is not integrable, but where $h(x) < f(x) < g(x)$ and both $h(x)$ and $g(x)$ *are* integrable. What is the significance of such an example?
Examples	Construct a pair of convergent series, $\sum u_n$ and $\sum v_n$, for which the ratio test is unable to show that $\sum v_n$ is convergent given that $\sum u_n$ is, and vice versa. Do a pair of convergent series exist for which the ratio test can used to prove the other convergent, but not vice versa? Construct an example that shows why *f* has to be decreasing in order that, 'For *f* non-negative, $\int_1^{\infty} f(x)\,dx$ exists iff $\sum_1^{\infty} f(n)$ exists.' is a theorem.

Bibliography

Abel, N. H. (1828) *Note sur le Mémoire de M.L.Olivier No.4 du Second Tome de ce Journal, Ayant pour Titre 'Remarques sur les Séries Infinies et leur Convergence'*, Crelle's Journal, 3 (1).

Anderson, J. Austin, K. Barnard, T. and Jagger, J. (1998) Do Third-year Mathematics Undergraduates Know What They are Supposed to Know?, *International Journal of Mathematical Education in Science and Technology*, 29 (3), p401–420. Abbreviated in TALUM Newsletter No. 8 (see *Websites*).

Angelo, T. and Cross, K. (1993) *Classroom Assessment Techniques*, 2nd Edition, Jossey-Bass, San Francisco.

Avital, S. and Barbeau, E. (1991) Intuitively Misconceived Solutions to Problems, *For the Learning of Mathematics*, 11 (3), p2–7.

Baker, J. Crampin, M. and Nuttall, J. (1973) A Crash Course in Calculus, *International Journal of Mathematics Education in Science and Technology*, 4, p335–339.

Bauersfeld, H. (1994) Theoretical Perspectives on Interaction in the Mathematics Classroom, in Biehler. R. *et al.* (eds) *The Didactics of Mathematics as a Scientific Discipline*, Kluwer, Dordrecht.

Barry, M. Williams, J. Wilson (2000) Computer Testing Using TAL, *Mathematics, Statistics, and Operations Research: Supporting Learning and Teaching*, third newsletter, August, p20–21.

Baumslag, B. (2000) *Fundamentals of Teaching Mathematics at University Level*, Imperial College Press, London.

Bills, E. (1996) Shifting Sands: Students' understanding of the roles of variables in 'A' level mathematics, unpublished PhD Thesis, Open University.

Boelkins, M. and Ratliff, T. (2001) How We Get Our Students to Read the Text Before Class, *Focus: The Newsletter of the Mathematical Association of America*, 21 (1), p16–17.

Boas, R. (1960) *A Primer of Real Function*, Mathematical Association of America, Washington.

Brousseau, G. (1984) The Crucial Role of the Didactical Contract in the Analysis and Construction of Situations in Teaching and Learning Mathematics, in Steiner, H. (ed.) *Theory of Mathematics Education*, Paper 54, Institut fur Didaktik der Mathematik der Universitat Bielefeld, p110–119.

Brousseau, G. (1997). *Theory of Didactical Situations in Mathematics: Didactiques des Mathématiques, 1970-1990*, Balacheff, N. Cooper, M. Sutherland, R. and Warfield, V. (trans.), Kluwer, Dordrecht.

Budden, F. (1972) *The Fascination of Groups*, Cambridge University Press, Cambridge.

Burn, B. (1996) What are the fundamental concepts of Group Theory?, *Educational Studies in Mathematics*, 31, p371–377.

Burn, B. (1998) Participating in the learning of Group Theory, *PRIMUS*, 8, p305–316.

Burn, R. Appleby, J. and Maher, P. (1997) *Teaching Undergraduate Mathematics*, Imperial College Press, London.

Casselman, B. (2000) *Pictures and Proofs*, Notices of the American Mathematical Society, 27 (10), p1257–1266.

Cornu, B. On Limits, in D. Tall (ed.) *Advanced Mathematical Thinking*, Kluwer, Dordrecht.

Courant, R. (1981) Reminiscences from Hilbert's Gottingen, *Mathematical Intelligencer*, 3 (4), p154–164.

Davis, J. (1987) Teaching by Correspondence in Mathematics, in Thorpe, M. and Grugeon, D. (eds) *Open Learning for Adults*, Longman Open Learning, Harlow.

Davis, P. (1993) Visual Theorems, *Educational Studies in Mathematics*, 24 (4), p333–337.

Dubinsky, E. (1991) Reflective Abstraction in Mathematical Thinking, in Tall, D. (ed.) *Advanced Mathematical Thinking*, Kluwer, Dordrecht.

Dubinsky, E., Dautermann, J., Leron, U. and Zazkis, R. (1994) On Learning Fundamental Concepts of Group Theory, *Educational Studies in Mathematics*, 27, p267–305.

Dubinsky, E., Dautermann, J., Leron, U. and Zazkis, R. (1997) A Reaction to Burn's 'What are the Fundamental Concepts of Group Theory?', *Educational Studies in Mathematics*, 34, p249–253.

Dubinsky, Ed. Schoenfeld. A. Kaput, Jim. (eds) (1994) *Research in Collegiate Mathematics Education I*, CBMS Issues in Mathematics Education, Vol. 4, American Mathematical Society, Providence.

Kaput, J. Schoenfeld, A. Dubinsky, E. (eds) (1996), *Research in Collegiate Mathematics Education II*, CBMS Issues in Mathematics Education, Vol. 6, American Mathematical Society, Providence. Reviewed in TALUM Newsletter No. 8 (see *Websites*).

Schoenfeld, A. Kaput, J., Dubinsky, E. (eds) (1998) *Research in Collegiate Mathematics Education III*, CBMS Issues in Mathematics Education, Vol. 7, American Mathematical Society, Providence. Reviewed in TALUM Newsletter No. 11 (see *Websites*).

Schoenfeld, A. Kaput, J., Dubinsky, E. (eds) (2000), *Research in Collegiate Mathematics Education IV*, CBMS Issues in Mathematics Education, Vol. 8, American Mathematical Society, Providence.

Feynman, R. (1995) *Six Easy Pieces: Essentials of physics explained by its most brilliant teacher*, Addison-Wesley, Reading.

Fischbein, E. (1987) *Intuition in Science and Mathematics: An educational approach*, Reidel, Dordecht.

Fischbein, E. (1993) Figural Concepts, *Educational Studies in Mathematics*, 24 (2) p139-162.

Flores, A. (1998) Galileo's Ratios, *The College Mathematics Journal*, 29 (4), p300.

Gold, B. (1999) *Assessing Mathematics*, Mathematical Association of America, Washington.

Gattegno, C. (1987) The Science of Education Part I: Theoretical considerations, Educational Solutions, New York.

Gavosto, E. Krantz, S. and McCallum, W. (1999) *Contemporary Issues in Mathematics Education*, Mathematical Sciences Research Institute Publications 36, Cambridge University Press, Cambridge.

Gibbs, G. (1992) *Improving the Quality of Student Learning*, Oxford Centre for Staff Development, Oxford.

Goar, M. (1999) Olivier and Abel on Series Convergence: An episode from 19th century analysis, *Mathematics Magazine*, 72 (5), p347–355.

Gulberg, J. (1997) *Mathematics From The Birth of Numbers*, Norton, New York.

Gray, E. and Tall, D. (1994) Duality, Ambiguity, and Flexibility: A proceptual view of simple arithmetic, *Journal for Research in Mathematics Education*, 25 (2), p116–140.

Harel, G. and Sowder, L. (1998) Student Proof Schemes: Results from exploratory studies, in Schoenfeld, Kaput , J. Dubinsky, E. (eds) *Research in Collegiate Mathematics Education III*, CBMS Issues in Mathematics Education, Vol. 7, p234–283.

Halmos, P. (1985) *I Want to be a Mathematician*, Springer, Berlin.

Halmos, P. (1994) What is Teaching?, *American Mathematical Monthly*, 101 (9), p848–854.

Halmos, P. (1980) The Heart of Mathematics, *American Mathematical Monthly*, 87 (7), p519–524.

Halmos, P. Moise, E. Piranian, G. (1975) The Problem of Learning to Teach, *American Mathematical Monthly*, 82 (5), p466–476.

Hannah, G. (1995) Changes to the Importance of Proof, *For the Learning of Mathematics*, 15 (3), p42–50.

Hawkes, T. and Savage, M. (2000) *Measuring the Mathematics Problem*, Engineering Council, London.

Hazzan, O. and Leron, U. (1996) Students' Use and Misuse of Mathematical Theorems: The case of Lagrange's Theorem, *For the Learning of Mathematics*, 16, p23–26.

Hewitt, D. (1996) Mathematical Fluency: The nature of practice and the role of subordination, *For the Learning of Mathematics*, 16 (2), p28–35.

Houston, K. and Lazenbatt, A. (in press) 'Peer Tutoring in a Modelling Course', *Innovations in Education and Training International.*

Houston, K. and Lazenbatt, A. (1996a) The Introduction and Evaluation of Peer Tutoring in Undergraduate Courses, *Journal of Further and Higher Education*, 20, p39–50.

Houston, K. and Lazenbatt, A. (1996b) A Peer-tutoring Scheme to Support Independent Learning and Group Project Work in Mathematics, *Assessment and Evaluation in Higher Education*, 21, p251–266.

Jackson, D. Fiddes, D. Beevers, C. (2000) Assessment Over The Web Using CUE, *Mathematics, Statistics, and Operations Research: Supporting Learning and Teaching*, third newsletter, August, p17–20.

Kilpatrick, J. (1981) Reasonable Ineffectiveness of Research in Mathematics Education, *For the Learning of Mathematics*, 2, p22–28.

Klein, J. (1934) *Greek Mathematical Thought and The Origin of Algebra*, Brann, E. (trans. 1992), Dover, New York.

Krantz, S. (1999) *How to Teach Mathematics* (second edition), Mathematical Association of America, Washington D.C.

Lakatos, I. (1976) *Proofs and Refutations*, Cambridge University Press, Cambridge.

Legrand, M. (1993) *Débate Scientifique en Cour de Mathématiques*, Repères IREM, No. 10, Topiques Edition.

Leikin, R. and Winicki-Landman, G. (2000) On Equivalent and Non-equivalent Definitions (part 2), *For the Learning of Mathematics*, 20 (2), p24–29.

Leron, U. and Dubinsky, E. (1995) An Abstract Algebra Story, *American Mathematical Monthly*, 102 (3), p227–242.

MAA Videotape (1966) *Challenge in the Classroom: the Methods of R. L. Moore*, MAA video Classics 3, Mathematical Association of America, Committee on Educational Media, Washington.

MacHale, D. (1980) The Predictability of Counterexamples, *American Mathematical Monthly*, 87, p752.

Marton, F. and Saljo, R. (1976) On Qualitative Differences in Learning, *British Journal of Educational Psychology*, 46, p4–11.

Mason , J. (1998) *Learning and Doing Mathematics* (second revised edition), QED Books, York.

Mason, J. (2000) Asking Mathematical Questions Mathematically, *International Journal of Mathematical Education in Science and Technology*, 31 (1), p97–111.

Mason, J. (2001) (in press) Mathematical Teaching Practices at Tertiary Level: Working group report, in Holton, D. (ed.), *The Teaching and Learning of Mathematics at University Level: An ICMI study*, Kluwer, Dordrecht.

Mason, J. Burton L. and Stacey K. (1982) *Thinking Mathematically* (third edition 1998), Addison Wesley, London.

Meier, J. and Rishel, T. (1998) *Writing in the Teaching and Learning of Mathematics*, MAA Notes 48, Mathematical Association of America, Washington.

Michener, E. (1978) Understanding Understanding Mathematics, *Cognitive Science*, 2, p361–383.

Moshovits-Hadar, N. (1988) Surprise, *For the Learning Of Mathematics*, 8 (3) p34–40.

Nardi E. (1992) Conceptions of Advanced Mathematical Concepts and Some Cognitive Phenomena Relating to Them: Difficulties, obstacles and errors; the exemplary case of function, in Gagatsis, A. (ed.) *Didactics of Mathematics*, Erasmus, Thessaloniki.

Nardi E. (1996) The Novice Mathematician's Encounter With Mathematical Abstraction: The case of accumulation point, in Gagatsis, A. and Rogers, L. (eds) *Didactics and History of Mathematics*, Thessaloniki, p227–235.

Nardi E. (1998) Mapping Out the First-year Mathematics Undergraduate's Difficulties with Abstraction in Advanced Mathematical Thinking: Didactical observations related to the teaching of advanced mathematics from a cross-topical study, in *Proceedings of the International Conference on the Teaching of Mathematics,* John Wiley and Sons, Pythagorion, p224–226.

Nardi, E. (1999) The Novice Mathematician's Encounter with Formal Mathematical Reasoning: A sample of findings from a cross-topical study, Oxford University, *TALUM Newsletter No. 9* (see *Websites*).

Nardi. E. (1999a) The Challenge of Teaching First-Year Undergraduate Mathematics: Tutors' reflections on the formal mathematical enculturation of their students, *Nordic Studies in Mathematics Education,* 7 (2), p29–53.

Nardi, E. (2000) Zeros and Ones in Advanced Mathematics: Transcending the Intimacy of Number, *For the Learning of Mathematics,* 20(3), p44-51.

Nardi, E. (2000a) The Concept of Supremum/Infimum of a Set: A problematic overture to the concept of limit?, in Jaworski, B. (ed.) *Proceedings of Day Conferences at The University of Exeter (Feb 2000) and The University of Loughborough (May 2000),* British Society for Research into Learning Mathematics, Department of Education, University of Oxford, Oxford.

Nardi (2000b) Mathematics Undergraduates' Responses to Semantic Abbreviations, 'Geometric' Images and Multi-level Abstractions in Group Theory, *Educational Studies in Mathematics,* 43, p169–189.

Nelson, R. (1993) *Proofs Without Words,* Mathematical Association of America, Washington.

Nelson, R. (2000) *Proofs Without Words II,* Mathematical Association of America, Washington.

Nering, E. (1970) *Linear Algebra and Matrix Theory,* Wiley, New York.

Nicholson, J. (1993) The Development and Understanding of the Concept of Quotient Group, *Historia Mathematica,* 20, p68–88.

Pimm, D. (1987) *Speaking Mathematically,* Hodder and Stoughton, London.

Pirie, S and Kieren, T. (1994) Growth in Mathematical Understanding: How can we characterise it and how can we represent it?, *Educational Studies in Mathematics,* 26 (2-3), p165–190.

Polya, G. (1957) *How to Solve it* (second edition), Princeton University Press, Princeton.

Polya, G. (1962) *Mathematical Discovery: On understanding, learning, and teaching problem solving* (combined edition), Wiley, New York.

Ramsden, P. and Entwistle, N. (1985) Student Learning Research: Retrospect and prospect, Higher Education Research and Development, 4, p51–69.

Ridley, J. (2000) An Elementary Proof on Location of Zeros, *American Mathematical Monthly,* 107 (4), p357–360.

Rosenbrock, H. (1960) An Automatic Method for Finding the Greatest or Least Value of a Function, *Computer Journal,* 3, p175–184.

Seldon, A. Seldon, J. Hauk, S. and Mason, A. (2000) 'Why can't calculus students access their knowledge to solve nonroutine problems?' in *Collegiate Mathematics Education IV*, CBMS Issues in Mathematics Education, Vol. 8, American Mathematical Society, Providence, p128–153.

van Rossum, E.J. and Schenk, S.M. (1984) The Relationship Between Learning Conception and Study Strategy and Learning Outcome, *British Journal of Educational Psychology*, 54, p73–83.

Schoenfeld, A. (1985) *Mathematical Problem Solving*, Academic Press, New York.

Schmidt, H.G. (1994) Special Issue on Problem-based Learning, *Instructional Science*, 22 (4), p247–340.

Sfard, A. (1991) On the Dual Nature of Mathematical Conceptions: Reflections on processes and objects as different sides of the same coin, *Educational Studies in Mathematics*, 22, p1–36.

Sfard, A. (1992) Operational Origins of Mathematical Notions and the Quandry of Reification: The case of function, in E. Dubinsky, and G. Harel (eds), *The Concept of Function: Aspects of epistemology and pedagogy*, MAA Monograph series, Washington.

Sfard, A. (1994) Reification as the Birth of Metaphor, *For the Learning of Mathematics*, 14 (1) p44–55.

Sierpinska, A. (1994) *Understanding in Mathematics*, Falmer Press, London.

Skemp, R. (1976) Relational Understanding and Instrumental Understanding, *Mathematics Teaching*, 77 (December), p20–26.

Smith, D. and Moore, L. (1991) The Laboratory Approach to Teaching Calculus, *MAA Notes*, 20.

Solow, D. (1990) *How to Read and Do Proofs: An introduction to mathematical thought processes*, Wiley, Chichester.

Sterrett, A. (undated) Using Writing to Teach Mathematics, *MAA Notes*, 16, Mathematical Association of America, Washington.

Sutherland, R. and Dewhurst, H. (1999) *Mathematics Education Framework for Progression from 16–19 to HE*, Graduate School of Education, University of Bristol, Bristol.

Szydlik, J. (2000) Mathematical Belleta and Conceptual Understanding of the Limit of a Function, *Journal for Research in Mathematics Education*, 31 (3), p258–276.

Tahta, D. (1981) Some Thoughts Arising from the New Nicolet Films, *Mathematics Teaching*, 94, p25–29. Reprinted in Beeney, R. Jarvis, M. Tahta, D. Warwick, J. and White, D. (1982) *Geometric Images*, Leapfrogs, Association of Teachers of Mathematics, Derby, p117–118.

Tall, D. (1992) The Transition to Advanced Mathematical Thinking: Functions, limits, infinity and proof, in D. Grouws (ed.) *Handbook of Research on Mathematics Teaching and Learning*, Macmillan, New York.

Tall, D. and Vinner, S. (1981) Concept Image and Concept Definition in Mathematics with Particular Reference to Limits and Continuity, *Educational Studies in Mathematics*, 12 (2), p151–169.

Tufte, E. (1983) *Envisioning Information*, Graphics Press, Cheshire.

Tufte, E. (1983) *The Visual Display of Quantitative Information*, Graphics Press, Cheshire.

Tufte, E. (1983) *Visual Explanations: Images and quantities, evidence and narrative*, Graphics Press, Cheshire.

van der Waerden, B. (1983) *Geometry and Algebra in Ancient Civilizations*, Springer-Verlag, Berlin.

Walker, J. (1975) *The Flying Circus of Physics with Answers*, Wiley, New York.

Wanner, E. and Hairer, G. (1996) *Analysis by its History*, Springer, New York.

Watson, A. and Mason, J. (1998) *Questions and Prompts for Mathematical Thinking*, ATM, Derby.

Wisner, R. (1999) *Powers as Uniform Sums of Positive Squares*, College Maths Journal, 30 (4), p293–296.

Wright, R. (1930) *Graduated Problems*, Cambridge University Press, Cambridge.

Wu, H. (1999) On the Education of Mathematics Majors, in Gavosto, E. Krantz, S. and McCallum, W. *Contemporary Issues in Mathematics Education*, Mathematical Sciences Research Institute Publications 36, Cambridge University Press, Cambridge, p9–23.

Yusof, Y. and Tall, D. (1994) Changing Attitudes to Mathematics Through Problem Solving, in da Ponte, J.P. and Matos, J.F. (eds) *Proceedings of the 18th Conference of the International Group for the Psychology of Mathematics Education*, Vol. IV, University of Lisbon, Lisbon, p401–409.

Historical Sources and Resources

Bos, H. (1993) *Lectures in the History of Mathematics*, History of Mathematics, Vol. 7, American Mathematical Society and London Mathematical Society, Providence.

Cajori, F. (1993) *A History of Mathematical Notations*, Dover, New York.

Calkin N. and Wilf, H. (2000) Recounting the Rationals, *American Mathematical Monthly*, 107 (4) p360–363.

Cambell, D. Higgins, J. (eds) (1984) *Mathematics: people, problems, results*, Wadsworth International, Belmont.

Chabert, J. (ed.) (1999) *A History of Algorithms: From the pebble to the microchip*, Springer, Berlin.

Dunham, W. (1999) *Euler: the Master of Us All*, Dolciani Mathematical Expositions, No. 22, Mathematical Association of America, Washington.

Fauvel, J. (2000) The Role of History of Mathematics within a University Mathematics Curriculum for the 21st Century, *TALUM Newsletter No. 11* (see *Websites*).

Fauvel, J. and Gray, J. (1987) *The History of Mathematics: A reader*, Macmillan Education, Basingstoke.

Fauvel, J. and van Maanen, J. (eds) (2000) *History in Mathematics Education: The ICMI study*, Kluwer, Dordrecht.

Katz, V. (1993) *A History of Mathematics*, Harper Collins, New York.

Katz, V. (2000) Using History to Teach Mathematics: An international perspective, *MAA Notes*, 51, Mathematical Association of America, Washington.

Kline, M. (1972) *Mathematical Thought from Ancient to Modern Times*, Oxford University Press, Oxford.

See also the series *History of Mathematics*, American Mathematical Society and London Mathematical Society, London.

Newman, J. (ed.) (1960) *The World Of Mathematics* (four volumes), George Allen & Unwin, London.

Smith, D. (1959) *A Source Book in Mathematics*, Dover, New York.

Swetz, F. Fauvel, J. Bekken, O. Johansson, B. and Katz, V. (eds) (1995) *Learn from the Masters*, Mathematical Association of America, Washington.

Tikhomirov, V. (1990) *Stories About Maxima and Minima*, Mathematical World Vol. 1, American Mathematical Society, Providence, Rhode Island.

Turnbull, H. (1962) *The Great Mathematicians*, Methuen, New York.

Wigner, E. (1960) The Unreasonable Effectiveness of Mathematics in the Natural Sciences, *Communications in Pure and Applied Mathematics*, 13, p1–14.

Wolff, P. (1963) *Breakthroughs in Mathematics*, Signet Science Library, New American Library, New York.

Teaching of Specific Mathematical Topics

Burn, R. (1992) *Numbers and Functions: Steps to analysis*, Cambridge University Press, Cambridge.

Carlson, D. Johnson, C. Lay, D. Porter, A. Watkins, A. and Watkins, W. (eds) Resources For Teaching Linear Algebra, *MAA Notes*, Vol. 42, Mathematical Association of America, Washington.

Dubinsky, E. and Harel, G. (eds) (1992) The Concept of Function: Aspects of epistemology and pedagogy, *MAA notes*, Vol. 25, Mathematical Association of America, Washington.

Dudley, U. (ed.) Resources For Calculus Collection, Volume 5: Readings for calculus, *MAA Notes*, Vol. 31, Mathematical Association of America, Washington.

Fraga, R. (ed.) (1999) Resources For Calculus Collection, Volume 2: Calculus problems for a new century, *MAA Notes*, Vol. 28, Mathematical Association of America, Washington.

Groetsch, C. (1999) *Inverse Problems: Activities for undergraduates*, Mathematical Association of America, Washington.

Jackson, M. & Ramsay, J. (eds) (1993) Resources For Calculus Collection, Volume 4: Applications of calculus, *MAA Notes*, Vol. 30, Mathematical Association of America, Washington.

Leinbach, L. Hundhausen, J. Ostebee, A. Senechal, L. and Small, D. (eds) (1991) The Laboratory Approach to Teaching Calculus, *MAA Notes*, 20, Mathematical Association of America, Washington.

Leron, U. (1983) Structuring Mathematical Proofs, *American Mathematical Monthly*, 90 (3), p174–184.

Leron, U. (1985) Heuristic Presentations: The role of structuring, *For the Learning of Mathematics*, 5 (3), p7–13.

Mason, J. Burton, L. and Stacey, K. (1982) *Thinking Mathematically*, Addison Wesley, London.

Schoenfeld, A. (1985), *Mathematical Problem Solving*, Academic Press, New York.

Schoenfeld, A. (1990) *A Source Book for College Mathematics Teaching*, Mathematical Association of America, Washington D.C.

Schoenfeld, A. (ed.) (1990) *A Source Book for College Mathematics Teaching*, Mathematical Association of America, Washington.

Solow, A. (ed.) (1993) Resources For Calculus Collection, Volume 1: Learning by Discovery: A lab manual for calculus, *MAA Notes*, Vol. 27, Mathematical Association of America, Washington.

Straffin, P. (ed.) (1993) Resources For Calculus Collection, Volume 3: Applications of calculus, *MAA Notes*, Vol. 29, Mathematical Association of America, Washington.

Classroom Techniques

Angelo, T. and Cross, K. (1993) *Classroom Assessment Techniques* (second edition), Jossey-Bass, San Francisco, p69–72.

Berry, J. and Sharp, J. (1999) Developing Student-centred Learning in Mathematics Through Co-operation, Reflection, and Discussion, *Teaching in Higher Education*, 4 (1), p27–41.

Hubbard, R. (1990) Tertiary Mathematics Without Lectures, *International Journal of Mathematical Education in Science and Technology*, 21 (4), p567–571.

Hubbard, R. (1991) *53 Interesting Ways to Teach Mathematics*, Technical and Educational Services, Oxford Brookes, Oxford.

Hubbard, R. (1995) *53 Ways to Ask Questions in Mathematics and Statistics*, Technical and Educational Services, Oxford Brookes, Oxford.

Lenker, S. (ed.) (1998) Exemplary Programs in Introductory College Mathematics: Innovative programs using technology, *MAA Notes*, Vol. 47, Mathematical Association of America, Washington.

Seymour, E. Hewitt, N. (1997) *Talking About Leaving: Why undergraduates leave the sciences*, Westview Press, Boulder.

Tucker, A. (ed.) (1995) Models That Work: Case studies in effective undergraduate mathematics programs, *MAA Notes*, Vol. 38, Mathematical Association of America, Washington.

Research into Teaching Mathematics at Tertiary Level

Artigue, M. (1998) De la Comprehension des Processus D'apprentissage a la Conception de Processus D'enseignement, in *Proceedings of the International Congress of Mathematicians,* Berlin 1998, Vol. III (Invited Lectures), Documenta Mathematica, Journal Der Deutschen Mathemtiker-Vereinigung, p723–733.

Dreyfus, T. (1991) On the Status of Visual Reasoning in Mathematics and Mathematics Education, Plenary Lecture, *PME XV,* Furinghetti, F. (ed.), Vol. I, p33–48, Assisi.

Dubinsky, E. (2000) Using a Theory of Learning in College Mathematics Courses, *TALUM Newsletter No. 12* (see *Websites*).

Michener, E. (1978) Understanding Understanding Mathematics, *Cognitive Science,* 2 p361–383.

Sfard, A. (1991) On the Dual Nature of Mathematical Conceptions: Reflections on processes and objects as different sides of the same coin, *Educational Studies in Mathematics,* 22, p1–36.

Sutherland, R. and Pozzi, S. (1995) *The Changing Mathematical Background of Undergraduate Engineers,* The Engineering Council.

Tall, D. (ed.) (1991) *Advanced Mathematical Thinking,* Kluwer, Dordrecht.

Perspectives on Mathematics

Halmos, P (1985) *I Want to be a Mathematician: An automathography,* Springer, Berlin.

Lakatos, I. (1976) *Proofs and Refutations,* Worral and Zahar (eds), Cambridge University Press, Cambridge.

Mason, J. Burton, L. and Stacey, K. (1982) *Thinking Mathematically* (third edition 1998), Addison-Wesley, London.

Polya, G. (1957) *How to Solve it,* Anchor, New York.

Soifer, A. (1987), *Mathematics as Problem Solving,* Centre for Excellence in Mathematical Education, University of Colorado, Colorado Springs.

Websites

Computer aided learning and assessment: http://www.icbl.hw.ac.uk/marble/maths/public (MARBLE), http://www.calm.hw.ac.uk (CALM) and http://www.tal.bris.ac.uk/ (TAL).

Common Mathematical Errors: http://www.math.vanderbilt.edu/~schectex/commerrs/

Glossary of Mathematical Mistakes (as seen in the media): http://www.mathmistakes.com

Teaching And Learning Undergraduate Mathematics newsletter (TALUM): http://www.bham.ac.uk/ctimath/talum/newsletter/

Metric project, Imperial College: http://metric.ma.ic.ac.uk/

Diagnosys at Newcastle: http://www.staff.ncl.ac.uk/john.appleby/diagpage/diagindx.htm

MacTutor History of Mathematics Archive:
http://www-groups.dcs.st-and.ac.uk/~history/

Earliest Uses of Various Mathematical Symbols:
http://members.aol.com/jeff570/mathsym.html

A Handbook for Mathematics Teaching Assistants:
http://www.maa.org/pfdev/tahandbook.html

American Mathematical Society links: http://www.ams.org/mathweb

Online mathematics books at Cornell:
http://digital.library.upenn.edu/webbin/book/subjectstart?QA

Proposed standards for courses bridging American High School and College:
http://www.imacc.org/standards/

Links to sites concerned with teaching mathematics at college level:
http://archives.math.utk.edu/materials.html

Discussion about forms of assessment in mathematics, further resources and links:
http://forum.swarthmore.edu/mathed/assessment.html
http://forum.swarthmore.edu/ library/

Warwick University also hosts an unmoderated web-discussion group devoted to issues in teaching mathematics at tertiary level. To get a copy of the guidelines, send an email to majordomo@warwick.ac.uk with 'get mathedu guidelines' as the subject line.

Index of Tactics, Issues, Themes, Frameworks and Tensions

Tactics

1 Some Student Difficulties with Mathematics **7**
Using Common Errors 10
Specialising, Generalising and Counter-Examples 11
Multiple Notations 13
Boundary Examples 14
Expressing Generality 15
Student Generated Exercises 16
Introducing Technical Terms 18
Using Multiple Representations 19
Say What You See 19
Inner Moves 21
Reconstructing and re-expressing 23
Introducing Definitions 24
Intensive and Extensive Definitions 25
Using 'Strange' Examples 25
Reconstructing Definitions 26
Distinguishing Common and Technical Meanings 27
Translating Technical Terms 28
Making an Example Be Exemplary 29
Say What You See (again) 30
Invariance Amid Change 32
Forging Links 33
Forcing Forces 34
Developing 'Physicacy' 35
Estimating Explicitly 36
Modelling Explicitly 37

2 Lecturing **39**
Treating 'Screens' as Phenomena 48
Advance Organisers and Summarising Organisers 48
Talking in Pairs 49
Pausing 50
Emphasising One Thing Per Session 50
Making Memorable Remarks 51
Muddiest and Most Important 51
Tracking The Intuitive and The Formal 52

Presenting the Essence of a Theorem 52
Getting Students Used to the Idea of Proof 53
Commenting on Proofs 55
Invoking Mental Imagery 55
Being Explicit About The Enterprise 56
Exposing Inner Monologues 57
Being Human 57
Learning from Experience 58
Taping Yourself 59
Punctuating Your Material 60
Interrogating Your Own (Analogous) Experience 62
Varying Notation 63
Extending Core Material 66
Theme Exploitation 67
Student Mentoring 67
Tape–Frames 68
Computer-based Mathematics 68

3 Tutoring **71**
Assessing Degrees of Confidence 72
Voting on Conjectures 73
Generating Discussion 74
Anti-Funnelling 76
Making the most of a worked example or exercise 79
Story Telling 82
Arranging and Sorting 83
Linking Terms 84
Scrambled Correspondences 84
Diverting Attention 85
Exemplification 86
Being Mathematical 87
Simplifying and Complexifying 88
Particular – Peculiar – General 88
Doing and Undoing 89
Introducing Symbols 90
Directed – Prompted – Spontaneous 91
Say What You See (yet again) 91
Drawing Diagrams 92
Scrambled proof 94
Proof In Particular 94
Catching 'it' 95
Finding out how other people think 95

Advising students how to study a mathematics text 96
Advising students how to make the most of a worked example 97
Advising students how to make the most of an exercise 97
Advising students how to integrate a topic by weaving a story 98
Advising students how to learn how to learn (Learning Files) 98
Advising students how to recognise a problem type 99
Advising students how to gain mastery of a technique 99
Self-Selection 101

4 Constructing Tasks 105
Using Students to Learn About Students 110
Using Students' Reflections 111
Commenting on Student Scripts 112
Promoting self-checking 112
Characterising 116
Locating Inner Aspects of Tasks 132
Making a Task Appropriately Challenging 132
Labelling Behaviour *Not* People 133
Hares and Tortoises 134
Boundary Examples (again) 136

5 Marking and Commenting 141
Using Group Convenors 142
Collaborative Work 142
Rewarding Awareness of Processes 142
Being clear about the mark scheme 144
Establishing an atmosphere of mutual support 144
Exchanging scripts 144
Focusing on what is mathematical 145
Developing a language 145
Finding something positive to say 146
Selecting What To Mark 146
Providing model solutions 146
Summarising your observations 147
Providing a list of common errors or a 'corrected' sample student argument 147
Learning from students 153

6 Making Use of History 155
Pictures and Stories 157
Locating Earliest Known Uses of Symbols or Ideas 158
Exploring Controversy 158
Making Links 159

Using Original Material 159

Issues

Time – coverage and pace 164
View – serialist or holist 165
Challenge – insufficient or excessive 166
Implications of student and tutor expectations 166
Doing is not the same as construing 168
Knowing and Understanding 171
What is exemplary about an example? 173
The Place of Definitions, Theorems and Examples In Mathematical Exposition 175
Discussing mathematics 176
Dependent and Independent Learners 177
Responding to 'Why are we doing this?' 179
Developing Facility 180
Locating Surprise 181
Exploiting Surprise, Interest and Enthusiasm 183
Intriguing and Pointing The Way 184

Themes

Mathematical Powers 184
Shift of Attention 186
Invariance Amid Change 192
Doing and Undoing 193
Freedom and Constraint 193
Extending Meaning 193
Organising and Characterising 194

Frameworks

Manipulating – Getting-a-sense-of – Articulating 187
See – Experience – Master 189
Awareness, Motivation, and Behaviour 190
Concept Images 190

Tensions

Agenda and expectations 163

Made in the USA
Monee, IL
18 October 2026

66036211R10129